ARMAMENT ENGINEERING

a computer aided approach

H Peter

© Copyright 2003 H. Peter. All rights reserved.

No part of this publication may be reproduced, stored in a retrieval system, or transmitted, in any form or by any means, electronic, mechanical, photocopying, recording, or otherwise, without the written prior permission of the author.

National Library of Canada Cataloguing in Publication

Peter, H.
Armament engineering : a computer aided approach / H. Peter.
Includes bibliographical references and index.
ISBN 1-4120-0241-9
I. Title.

UG145.P47 2003 623.4'1 C2003-902440-7

TRAFFORD

This book was published *on-demand* in cooperation with Trafford Publishing. On-demand publishing is a unique process and service of making a book available for retail sale to the public taking advantage of on-demand manufacturing and Internet marketing. **On-demand publishing** includes promotions, retail sales, manufacturing, order fulfilment, accounting and collecting royalties on behalf of the author.

Suite 6E, 2333 Government St., Victoria, B.C. V8T 4P4, CANADA
Phone 250-383-6864 Toll-free 1-888-232-4444 (Canada & US)
Fax 250-383-6804 E-mail sales@trafford.com
Web site www.trafford.com TRAFFORD PUBLISHING IS A DIVISION OF TRAFFORD HOLDINGS LTD.
Trafford Catalogue #03-0610 www.trafford.com/robots/03-0610.html

10 9 8 7 6

Dedicated To

My Wife

And to

My Father

This book was

Electronically type set using Microsoft Word 2000 Professional

The body of the text was set in Century Schoolbook

The graphics drawn in Corel Draw Version 5.00.E2, Corel Corporation 1994

The equations set using Microsoft Equation Editor, Version 3.1, Microsoft Corporation

The spreadsheets created using Microsoft Excel 2000 Professional

The programmes were run in Matlab Version 5.2.0.3084 1988, The Mathworks Inc

and

The cover designed using Adobe Photoshop CS

by the author

Acknowledgements

The encouragement from family and friends made this book possible.

Introduction

It is expected that the treatment contained here will be equally useful to both students and instructors of Armament Engineering at colleges offering the field as a specialization.

This volume covers essential mechanical engineering aspects common to artillery and tank main armament. It does not deal with rocket launchers mortars or automatic anti aircraft cannon.

A base in mechanical engineering subjects; Strength of Materials, Fluid Mechanics, Thermodynamics, Applied Mechanics, Theory of Mechanisms, Machine Design and of course Engineering Mathematics as also exposure to Internal Ballistics is presumed.

Four key facets integrated here are Mechanical Engineering, Interior Ballistics, Armament Operation, and Computer Programming.

Computer programming allows immediate application of concepts learnt with unmatched flexibility when analysing the effect of variables. The softwares used Matlab and Excel, do not demand any previous programming skills.

This book aims to reach the student to the verge of armament design. Subsequently, in design, relevant aspects of interior ballistics and Armament Engineering are combined towards the investigation, analysis and solution of representative armament design exercises.

The Author

Contents

3 Recoil Systems

4 Muzzle Brakes

5
Supporting Structures

6
Elevating and Traversing Mechanisms

7
Balancing

Appendices

Index

List of Figures

1.6

1.7

1.8

2

2.1

2.2

3.6

3.7

3.8

3.9

4

4.1

4.2

4.3

5

5.1

5.2

5.3

6

6.2

6.3

7

7.1

7.2

7.3

1

Gun Barrels

1.1: Gun Barrels in General

Components of Artillery Weapons and Tank Main Armament

Artillery weapons and tank main armament consist of two main components; the gun itself and the supporting structure from which it fires. In the case of an armoured fighting vehicle (AFV), such as a tank or an artillery gun mounted in a turret on a tank chassis, also known as a self propelled mounting, the vehicle hull is the supporting structure. The gun is traversed by the rotation of the turret on the hull. In the case of towed artillery, the supporting structure is called the carriage when the gun fires with the wheels in ground contact. When the gun fires off rigid supports, the wheels being taken out of contact with the ground after transportation, the platform is generally called a mounting. This terminology is not universal, but of material importance are the components and the functions of the supporting structure. Whether called a carriage or a mounting, all supporting structures perform certain common and important functions.

Supporting Structure

The supporting structure carries the gun and ensures stability to it especially during and immediately after firing when reaction forces act. The supporting structure includes the arrangements for aiming the gun and facilitates movement of the gun from place to place by doubling as a vehicular chassis, which is attached to a prime mover. Supporting structures may also have an integral means of propulsion for tactical movement within a limited radius.

The Gun

The gun itself consists of the following main parts:

- The barrel.
- Barrel attachments, like muzzle brakes, fume extractors, flash hiders etc.
- The breech assembly, which includes the firing mechanism.

The gun itself rests on a component called the cradle. The cradle acquires it name from its similarity to the infant crib. When the cradle is rocked about its transverse axis, the pivots being called the trunnions, the gun also gets rocked in the vertical plane. The gun is allowed movement in its axial direction by means of slides, attached to the gun, moving on longitudinal guides, which are integral to the cradle. The guides are an essential feature of all cradles. The gun is linked flexibly to the cradle by means of a braking assembly called the recoil system, which retards motion of the gun during recoil.

The Barrel

The barrel or gun tube, as it is often referred to, is basically a high-pressure pressure vessel in which the chemical energy of the propellant is converted to the energy of translation of the projectile in a safe and consistent manner. The barrel also has the function of imparting direction and the necessary angle of elevation to the projectile. The barrel is closed at the breech end by the breech assembly and sealed temporarily towards the muzzle end by the projectile while it is in the barrel. The projectile performs a similar role as that of a piston in an internal combustion engine.

A barrel is commonly divided into four regions:

- The rear opening or breech.
- The chamber, which houses the complete round before firing
- The elongated portion or guidance section through which the projectile travels on firing. The inner surface of the guidance section is commonly called the bore. In the case of projectiles stabilized by spin, a system of grooves is cut on the inner surface of the guidance section. This is known as rifling. Spin is imparted to the projectile by interaction of the rifling with the ductile material of the driving band on the projectile by a process called engraving of the driving band. During its travel up the bore, a small portion of the translational energy of the projectile is converted to its rotational motion.
- The forward opening or muzzle.

For reasons of mobility and economy, a gun barrel should be as light as practically possible. The barrel should be sufficiently strong towards the chamber end to withstand the stresses induced by propellant gas pressure. Also as gun barrels are invariably supported towards the breech end, the barrel has to be rigid enough so that the deflection of the axis of the bore from its intended axis, due to its own weight should be within acceptable close limits. Towards the muzzle end, stress due to gas pressure is not the deciding factor. Here the barrel needs to be sufficiently thick and strong to bear the weight of muzzle attachments, also to cater for weakness caused by cutting of threads for their attachment.

Materials for Gun Barrels

Early gun barrels were cast from iron or bronze, for then, casting of gun barrels was the order of the day. For the muzzle velocities in evidence at the time, such metals sufficed. Currently, in accordance with high muzzle velocities, gun barrels are always made of steel, an alloy of iron with carbon in varying proportion. The carbon content in gun steels lies between 0.2 to 0.5%. Nickel, Chromium, Molybdenum and Silicon are added to enhance properties such as durability and rigidity, while Sulphur and Phosphorous are avoided due to their detrimental effects, like corrosion, on barrel life.

Components of a Gun Barrel

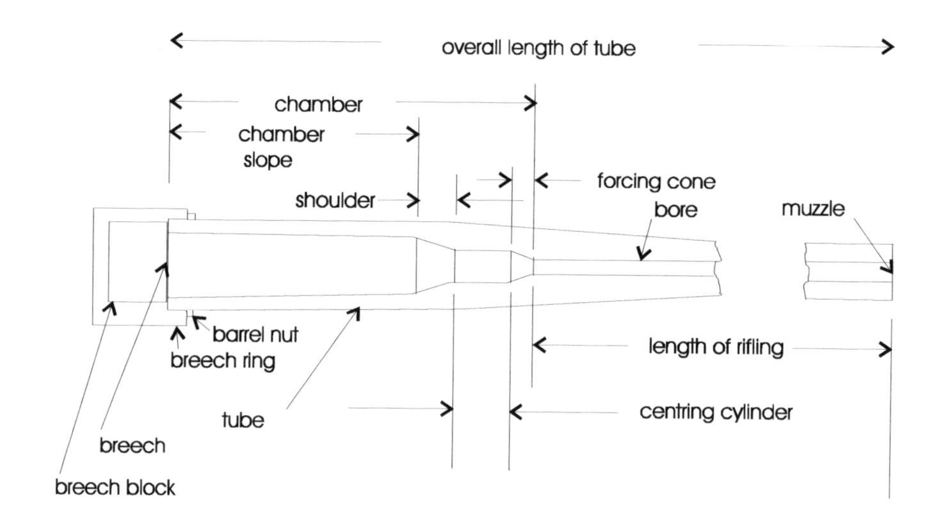

Fig 1.1.1: Components of a gun barrel

Breech and Breech Assembly

A sliding breechblock or a breech screw, which in both cases is supported by a frame called a breech ring, closes the breech end of the chamber. The breech ring is threaded on to the outside of the chamber of the barrel and secured to it by a barrel nut with necessary anti-rotation devices to prevent rotation of the barrel within the breech ring. On firing the accelerating component of the propellant gas force is transmitted to the base of the cartridge case then to the breechblock and ultimately to the breech ring. Consequently the breechblock, or screw and breech ring are heavily stressed during firing.

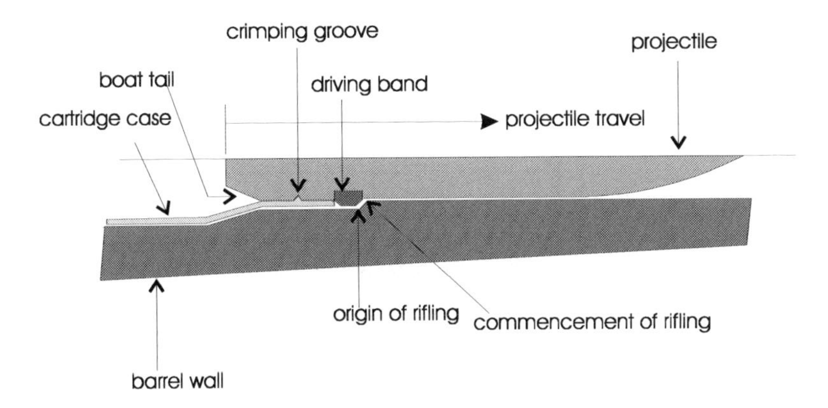

Fig 1.1.2: Relative position of cartridge case, driving band and rifling before firing

Chamber

This is the smooth portion of the interior of the barrel towards the breech end in which combustion of the propellant takes place. The chamber dimensions are related to the volume necessary for efficient combustion of the propellant as also the overall length of the barrel. Two kinds of chambers are common. In the bag loaded charge system, the rear end of the chamber is tapered to seat the obturator. In the QF popularly known as quick firing or metallic cartridge case charge system, the chamber is sloped to facilitate easy extraction of the empty cartridge case. On loading, the complete round rests in the chamber. Fig 1.1.2 depicts the relative position of the cartridge case, projectile, driving band and rifling before firing.

QF Chambers

In QF or cartridge case charge systems, the chamber and cartridge case profile are intimately associated. The chamber consists of the regions enumerated on the next page.

- Chamber body. This is the main part of the chamber; it houses the propellant and primer. In the case of QF systems, the cartridge case and the barrel expand elastically during firing, the cartridge case being of more ductile material than that of the barrel, expands by a greater amount, to provide a tight seal between the chamber wall and the lips of the case itself. After the pressure drops, the cartridge case contracts, more than the barrel material, again for the same reason of greater ductility, affording the necessary clearance between the chamber wall and the cartridge case itself for easy extraction of the spent case.

- Shoulders. These are the tapered portions of the chamber by means of which the chamber diameter is reduced to the bore diameter. The chamber volume is fixed by internal ballistics considerations. In order to limit the length of the chamber, hence the overall length of the gun, for a given chamber volume, the chamber diameter is usually greater than that of the bore. Depending on the ratio of chamber body diameter to bore diameter involved, one or more shoulders may be necessary. The chamber diameter is reduced smoothly to that of the bore. The creation of abrupt surfaces, which adversely affect the internal ballistics and life of the barrel is avoided.

- Centering cylinder. The diameter of the centering cylinder is equal to that of the driving band. Here the longer axis of the projectile and the axis of the bore are brought into alignment when the projectile is loaded.

- Forcing cone. This is a conical frustum whose slope extends through the origin of rifling and intersects the bore surface. The origin of rifling is the point at which the grooves begin and the commencement of rifling pertains to the point where the grooves attain maximum and thereafter constant depth. Engraving of the driving band occurs in the forcing cone. This region is hence highly stressed and more prone to wear than other parts of the bore.

BL Chambers

In the case of the bag-loaded charge system, a resilient obturator pad expands to seal the gap between the breech ring and screw, thereby providing the gas sealing. In such systems, based on bag charges, the chamber falls into three distinct parts:

- The coned portion designed to seat the obturator
- Main cylindrical body of the chamber
- The shot seating

Bore

The bore is the inner surface of the barrel cylinder in which the projectile is imparted acceleration. Because spinning an elongated projectile about its longer axis increases its stability; the inner surface of the bore is rifled in guns where the method of stabilization by spin is used. In smooth bore guns, the projectile is stabilized by fin stabilization. Rifled guns may also fire modified fin-stabilized projectiles. In such cases, a slipping or rotating driving band is incorporated in the projectile. Here the driving band provides the necessary gas sealing but does not transfer rotational motion to the projectile. The diameter of the projectile is always less than the caliber or land-to-land bore diameter. The diameter of the driving band or bands is naturally greater than that of the bore.

Types of Gun Barrels

Gun barrels are usually classified according to the method involved in their manufacture. Historically a number of ingenious methods have been practiced in keeping with the limitations of metallurgical and manufacturing techniques then prevalent. Currently the trend is towards monobloc and monobloc autofrettaged barrels. Analysis of some of the rather obsolete methods of barrel construction, especially the built up or shrink fitted barrel, affords useful information.

Loose Barrel/Loose Liner Barrel

In loose barrel or loose liner barrels, a jacket was fitted over part of the barrel, to provide strength to the highly stressed portions towards the breech end, as also to provide girder strength to the barrel along its length. In such barrels, the liner was easily removable for replacement.

Composite Barrels

In the composite barrel, individual segments of the barrel were fabricated from different grades of steel depending on the propellant gas pressures involved. Individual segments of these barrels were replaceable, but gas sealing between segments proved a problem with this type of design.

Monobloc Barrels

Monobloc barrels are homogeneous one piece metal forging. As such they are easy, quick and economical to manufacture. Modern alloy steels are suitable for monobloc barrels in which the pressures experienced are comparatively moderate. Monobloc barrel manufacture is a rather straightforward process. Presently available computerized numeric controlled machine tools are capable of mass-producing monobloc barrels economically, and to stringent quality control standards. Specialized processes like impressing the rifling pattern can be included in the auto-forging process. This is a time saving improvement over the method of cutting the rifling grooves. Here the grooves and land profile is forged onto the bore surface by insertion of a mandrel, which conforms to the groove-land profile, into the tube, which is of slightly larger diameter than that of the finished bore. The material of the barrel is then forged onto the mandrel profile by hammering.

The main drawback of monobloc construction is that when very high gas pressures are demanded, as in modern tank guns, with kinetic energy projectiles as their primary ammunition, the stresses in the barrel material exceeds the yield point of steel in use, and require a higher grade of steel or some form of pre-stressing. This has to be incorporated in the manufacturing process. Further, in monobloc barrels, the outer layers of the gun barrel are not stressed to the extent which justifies the cost of the high-grade steel used. From the stress point of view, the yield strength of steel required at the inner layers of the barrel wall is higher than towards the outer layers. This results in uneconomical use of high-grade steels. The suitability of monobloc barrels is directly related to the strength of the steel used. This point will be amplified later in this chapter in the section on monobloc barrels.

Wire Wound Barrels

In wire wound gun barrels, tensioned wire is wound in layers over the chamber of a steel tube, a uniform tension being maintained for each layer. Each layer exerts compression on the layer beneath it and consequently on the steel tube. Finally a jacket is shrunk on to the assembly. Wire can be made of tremendous strength. For this reason, wire wound barrels are stronger than barrels made by any other method within a given weight limit. In built up barrels, the tensile hoop stress exerted by the inner tube on the outer is a serious disadvantage. In wire wound barrels, the layers of wire apply only inward radial pressure; therefore the maximum stressed region remains at the inner surface of the tube. Wire wound barrels however lack girder strength and pose problems during the manufacturing process.

Built Up Barrels

This is now a more or less outdated form of pre-stressing, though the technique is used when it is desired to provide a hard liner in barrels designed for high rates of fire. In the built up method, multiple barrels were heat shrunk onto each other. In the simplest form of built up barrel, an outer steel tube of slightly lesser inner diameter than the outer diameter of an inner tube was heated. The outer tube was expanded by heating and fitted over the inner tube in the expanded or hot condition. On cooling it shrunk onto the inner tube, inducing compressive stress in the inner tube. The outer tube was also stressed in tension as it was prevented from shrinking to its original inner diameter by the presence of the inner tube. Hence from an initially unstressed steel, a compound tube was obtained with the inner tube under compressive hoop stress. This compressive stress had to be neutralized by the tensile hoop stress caused by the propellant gas pressure before tensile hoop stress was actually experienced by the inner tube. The interference or the difference between the inner diameter of the outer tube and the outer diameter of the inner tube was critical because the post shrink fitting stresses had necessarily to be below the yield of the steel for this method to succeed. Multiple concentric tubes could be shrunk onto the inner tube provided the yield stress of the steel was nowhere exceeded. Built up barrels were generally lighter than monobloc barrels and also possessed greater girder strength due to the lamination of two or more layers over the inner tube. Built up barrel manufacture was expectedly costlier and more complicated than that of monobloc barrels.

Autofrettage

Autofrettage is a current form of pre stressing in vogue for high velocity gun tubes such as tank main guns. During the process, a single tube of internal diameter slightly less than the desired calibre is chosen. The tube is subjected to an internal pressure to enlarge the inner layers beyond the elastic limit, while the outer most layers are stressed within the elastic limit. On removal of the applied internal pressure, the inner layers remain permanently deformed, while the outer layers, being elastic, compress the inner layers, themselves being placed under tensile stress due to the plastic deformation of the inner region of the barrel wall. On being subject to propellant gas pressure, the compressive stress on the inner layers neutralizes some of the tensile stress caused by the internal pressure.

As a beneficial result of autofrettaging, a lower grade of steel may be used for a given peak gas pressure. Alternatively, for a given grade of steel; a lesser wall thickness is acceptable. Additional advantages of the process are that proofing is rendered unnecessary as also fissures in the inner surface tend to get closed due to compression, reducing the chances of crack propagation during firing.

1.2: General Equations for Stresses in Gun Barrels under Internal and External Pressure

Stress Analysis of Gun Barrels

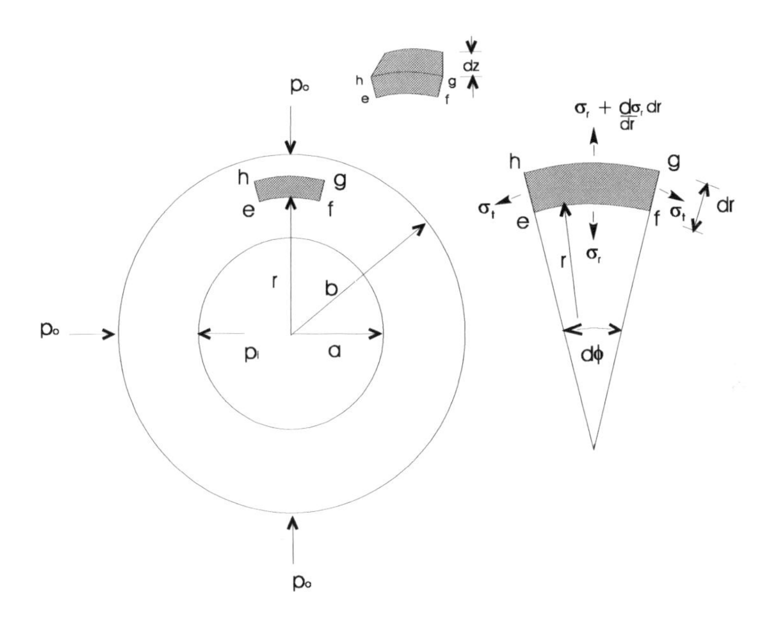

Fig 1.2.1: Cross section of a gun tube under symmetrical loading.

Stress analysis of gun barrels follows the methodology of stress analysis of thick walled tubes subject to pressure applied at the inner and outer surfaces, as also due to pressure within the material itself as a result of deformation. Combinations of internal and external pressure application as relevant to gun barrels will be analysed further in the text.

a: inner radius of the barrel
b: outer radius of the barrel
σ: stress
u: radial displacement
p_i: pressure at barrel inner surface
p_o: pressure at barrel outer surface
$d\varphi$: angle subtended by element at barrel center

Subscripts:

t: hoop stress
r: radial stress

The gun barrel may be considered a thick walled tube without restraint in the axial direction. The axial stress and strain is therefore insignificant. With reference to Fig 1.2.1, consider an element efgh of the gun barrel, of unit depth $dz = 1$ in the axial direction and thickness in the radial direction dr.

Let σ_t be the hoop stress acting normal to sides eh and fg.
Let σ_r be the radial stress acting normal to side ef.
This stress varies with the radius r and changes by an amount $\frac{d\sigma_r}{dr}dr$ over the distance dr.

Hence the radial stress on the side gh is: $\sigma_r + \frac{d\sigma_r}{dr}dr$

Summing up the forces in the direction of the bisector of the angle $d\varphi$ gives:

$$\sigma_r r d\varphi + \sigma_t dr d\varphi - \left(\sigma_r + \frac{d\sigma_r}{dr}dr\right)(r + dr)d\varphi = 0$$

Neglecting smaller quantities:

$$\sigma_t - \sigma_r - r\frac{d\sigma_r}{dr} = 0 \quad \text{[1.2.1]}$$

The two unknowns are the radial stress σ_r and the hoop stress σ_t.

Deformation of the Cylinder

If u denotes the radial displacement of a cylindrical surface of radius r the displacement for a surface of radius $r + dr$ is:

$$u + \frac{du}{dr}dr$$

The total elongation of the element efgh in the radial direction is:

$$\frac{du}{dr}dr$$

The unit radial elongation or radial strain is:

$$\varepsilon_r = \frac{u + \frac{du}{dr}dr - u}{dr}$$

$$\varepsilon_r = \frac{du}{dr}$$

Unit elongation of the circumference or the circumferential strain is:

$$\varepsilon_t = \frac{2\pi(r+u) - 2\pi r}{2\pi r}$$

Or:

$$\varepsilon_t = \frac{u}{r}$$

Equations for the Stresses in Terms of the Strains

Given that the Young's Modulus of Elasticity for the material is E and the Poisson's Ratio is μ the equations for the stresses in terms of the strains are:

$$\sigma_r = \frac{E}{(1-\mu^2)}\left(\frac{du}{dr} + \mu\frac{u}{r}\right) \qquad [1.2.2]$$

Differentiating Equation [1.2.2] with respect to r:

$$\frac{d\sigma_r}{dr} = \frac{E}{1-\mu^2}\left[r\frac{d^2u}{dr^2} + \mu\left(\frac{du}{dr} - \frac{u}{r}\right)\right] = \frac{E}{1-\mu^2}\left[\frac{d^2u}{dr^2} + \mu\left(\frac{1}{r}\frac{du}{dr} - \frac{u}{r^2}\right)\right]$$

$$\sigma_t = \frac{E}{(1-\mu^2)}\left(\frac{u}{r} + \mu\frac{du}{dr}\right) \qquad [1.2.3]$$

Substituting for $\sigma_t, \sigma_r,$ and $\frac{d\sigma_r}{dr}$ in Equation [1.2.1]

$$\frac{E}{(1-\mu^2)}\left(\frac{u}{r} + \mu\frac{du}{dr}\right) - \frac{E}{(1-\mu^2)}\left(\frac{du}{dr} + \mu\frac{u}{r}\right) - \frac{E}{1-\mu^2}\left[r\frac{d^2u}{dr^2} + \mu\left(\frac{du}{dr} - \frac{u}{r}\right)\right] = 0$$

Or:

$$\left(\frac{u}{r}+\mu\frac{du}{dr}\right)-\left(\frac{du}{dr}+\mu\frac{u}{r}\right)-\frac{E}{1-\mu^2}\left[r\frac{d^2u}{dr^2}+\mu\left(\frac{du}{dr}-\frac{u}{r}\right)\right]=0$$

Hence:

$$\frac{d^2u}{dr^2}+\frac{1}{r}\frac{du}{dr}-\frac{u}{r^2}=0 \quad \text{[1.2.4]}$$

The general solution of this second order differential equation is:

$$u=C_1r+\frac{C_2}{r} \quad \text{[1.2.5]}$$

Differentiating Equation [1.2.5]:

$$\frac{du}{dr}=C_1-\frac{C_2}{r^2} \quad \text{[1.2.5a]}$$

The constants C_1 & C_2 are determined from the conditions at the inner and outer surfaces of the cylinder where the normal stresses are known.

Substituting from Equations [1.2.5] and [1.2.5a] into Equations [1.2.2] and [1.2.3] respectively:

$$\sigma_r=\frac{E}{(1-\mu^2)}\left(C_1(1+\mu)-C_2\frac{1-\mu}{r^2}\right) \quad \text{[1.2.6]}$$

$$\sigma_t=\frac{E}{(1-\mu^2)}\left(C_1(1+\mu)+C_2\frac{1-\mu}{r^2}\right) \quad \text{[1.2.7]}$$

If p_i and p_o denote the pressures at the internal and external surfaces of radius a and b respectively, the stresses at the inner and outer surfaces of the barrel are:

$\sigma_{r\,inner}=-p_i$ and $\sigma_{r\,outer}=-p_o$.

The negative sign here denotes compressive stress.

Substituting $\sigma_{r\,inner} = -p_i$ *and* $\sigma_{r\,outer} = -p_0$ & $r = a$, $r = b$ in Equation [1.2.6]

$$C_1 = \frac{1-\mu}{E}\frac{a^2 p_i - b^2 p_o}{b^2 - a^2} \quad \text{[1.2.8]}$$

$$C_2 = \frac{1+\mu}{E}\frac{a^2 b^2 (p_i - p_o)}{b^2 - a^2} \quad \text{[1.2.9]}$$

With these values of C_1 and C_2, the general expressions for the stresses become:

$$\sigma_r = \frac{a^2 p_i - b^2 p_o}{b^2 - a^2} - \frac{(p_i - p_o)a^2 b^2}{r^2(b^2 - a^2)} \quad \text{[1.2.10]}$$

And:

$$\sigma_t = \frac{a^2 p_i - b^2 p_o}{b^2 - a^2} + \frac{(p_i - p_o)a^2 b^2}{r^2(b^2 - a^2)} \quad \text{[1.2.11]}$$

Gun Barrel Subject to Internal Pressure Alone

The condition of barrel in general subject to internal pressure is of interest when analyzing the stresses caused in barrels by the pressure of the propellant gas. In particular the analysis is applicable in the case of stress analysis of the outer tube in the case of built up barrels and in the case of the elastic region of autofrettaged tubes.

Considering the particular case when the external pressure p_o on the barrel = 0, i.e. when the barrel is subjected to internal pressure alone, then Equations [1.2.10] and [1.2.11] become:

$$\sigma_r = \frac{a^2 p_i}{b^2 - a^2}\left(1 - \frac{b^2}{r^2}\right) \qquad [1.2.12]$$

$$\sigma_t = \frac{a^2 p_i}{b^2 - a^2}\left(1 + \frac{b^2}{r^2}\right) \qquad [1.2.13]$$

The maximum value of σ_t is at the inner surface where $r = a$

$$\sigma_t = \frac{p_i}{b^2 - a^2}\left(a^2 + b^2\right)$$

The minimum value of σ_t is at the outer surface where $r = b$

$$\sigma_t = \frac{2a^2 p_i}{b^2 - a^2}$$

The shear stress is maximum at the inner surface where:

$$\tau_{\max} = \frac{\sigma_t - \sigma_r}{2} = \frac{p_i b^2}{b^2 - a^2}$$

Gun Barrel Subject to External Pressure Alone

The pressure on the outer surface of a gun barrel is always atmospheric and as compared to the high pressures within, is negligible. Hence this condition finds special application during the stress analysis of the inner tube of a built up barrel and also the plastic region of a tube subject to autofrettage.

In this case of the barrel subject to external pressure alone, as the internal pressure acting on the tube, $p_i = 0$, Equations [1.2.10] and [1.2.11] give:

$$\sigma_r = -\frac{p_o b^2}{b^2 - a^2}\left(1 - \frac{a^2}{r^2}\right) \qquad [1.2.14]$$

$$\sigma_t = -\frac{p_o b^2}{b^2 - a^2}\left(1 + \frac{a^2}{r^2}\right) \quad \text{[1.2.15]}$$

Here both stresses are apparently compressive in nature. The maximum compressive stress is at the inner surface:

$$\sigma_{t-inner} = -\frac{2p_o b^2}{b^2 - a^2}$$

Deformation of the Barrel

Substituting for C_1 and C_2 in Equations [1.2.5]

$$u = \frac{1-\mu}{E}\frac{a^2 p_i - b^2 p_o}{b^2 - a^2} r + \frac{1+\mu}{E}\frac{a^2 b^2 (p_i - p_o)}{(b^2 - a^2) r}$$

Radial displacement at the inner surface when the barrel is subject to internal pressure alone

$$u_{r=a} = \frac{a p_i}{E}\left(\frac{a^2 + b^2}{b^2 - a^2} + \mu\right) \quad \text{[1.2.16]}$$

Radial displacement at the outer surface when the barrel is subject to external pressure alone:

$$u_{r=b} = -\frac{b p_o}{E}\left(\frac{a^2 + b^2}{b^2 - a^2} - \mu\right) \quad \text{[1.2.17]}$$

The minus sign indicates contraction towards the axis of the barrel.

1.3: Monobloc Gun Barrels

Stresses in a Monobloc Barrel

From the stress analysis angle, a monobloc barrel is a cylindrical pressure vessel, of homogeneous material, subject to internal pressure alone. The pressure in this case is that of the propellant gas generated during firing. Equations [1.2.12] and [1.2.13] under conditions of internal pressure alone are applied directly.

Radial stress:

$$\sigma_r = \frac{a^2 p_i}{b^2 - a^2}\left(1 - \frac{b^2}{r^2}\right)$$

Hoop stress:

$$\sigma_t = \frac{a^2 p_i}{b^2 - a^2}\left(1 + \frac{b^2}{r^2}\right)$$

p_i: maximum gas pressure reached in the barrel.
a: internal radius of the barrel
b: external radius of the barrel
r: radius at any point intermediate between a and b

Example 1.3.1

A. The maximum gas pressure in a monobloc barrel is 20 tsi (UK). Compute and plot the hoop and radial stresses against the barrel thickness if the barrel is of internal diameter 100mm and thickness 50mm.
B. What is the effect on the stresses if the thickness of the barrel is increased by 25 mm?

Solution: Example 1.3.1 Part A

- In the first case, the stresses are computed with the help of Equations [1.1.12] and [1.2.13] for values of radius from 50 to 100mm and plotted against the same range of radius values.

- In the second case, the stresses are computed with the help of the same Equations [1.1.12] and [1.2.13] for radius values from 50 to 125mm and plotted from the same array of radius values.

Computer Programme 1.3.1 A

```
% programme to compute stresses in a monobloc barrel
a=.05 % inner radius
b=.100 % outer radius
p_i=20*1.544*10^7 % propellant gas pressure
r=(.05:.01:.1) % radius array
sigma_r=a.^2*p_i./(b.^2-a.^2)*(1-b.^2./r.^2) % radial stress
sigma_t=a.^2*p_i./(b.^2-a.^2)*(1+b.^2./r.^2) %hoop stress
plot(r,sigma_r,r,sigma_t) %plot stress vs radius.
```

Results: Example 1.3.1 Part A

The radial and hoop stress of Example 1.3.1 (A) are given below in tabulated form and the graphical depiction is contained in Fig 1.3.1 ahead.

Radius m	**Radial stress Pa*10^8**	**Hoop stress Pa*10^8**
0.0500	-3.0880	5.1467
0.0600	-1.8299	3.8886
0.0700	-1.0713	3.1300
0.0800	-0.5790	2.6377
0.0900	-0.2414	2.3001
0.1000	0	2.0587

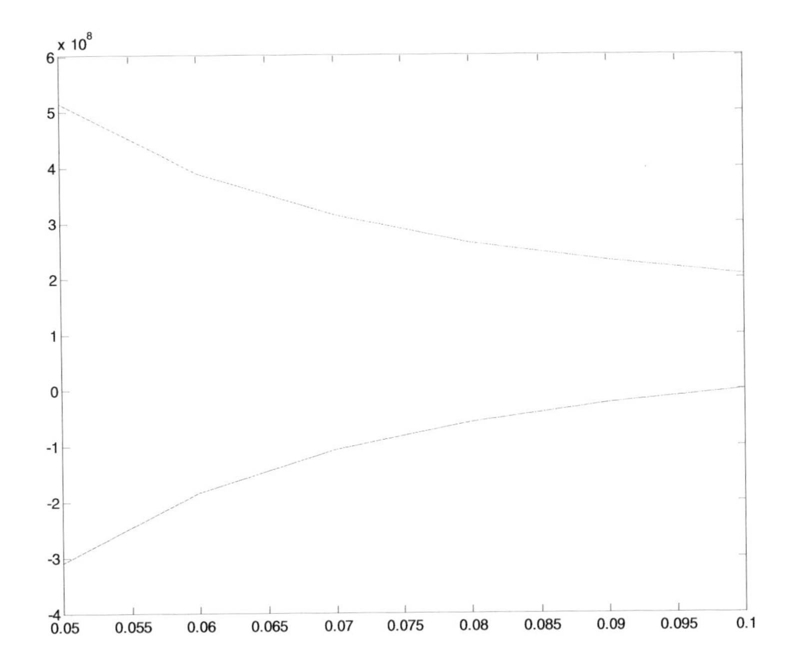

Fig 1.3.1: Hoop stresses in monobloc gun barrel of Example 1.3.1 A

Solution: Example 1.3.1 Part B

Computer Programme 1.3.1 B

The computer programme is essentially similar to that in the solution of Part A, but for the new value of the outer radius of the gun barrel.

Results: Example 1.3.1 Part B

Increase in the thickness results in a decrease in the stresses as seen from the plot below.

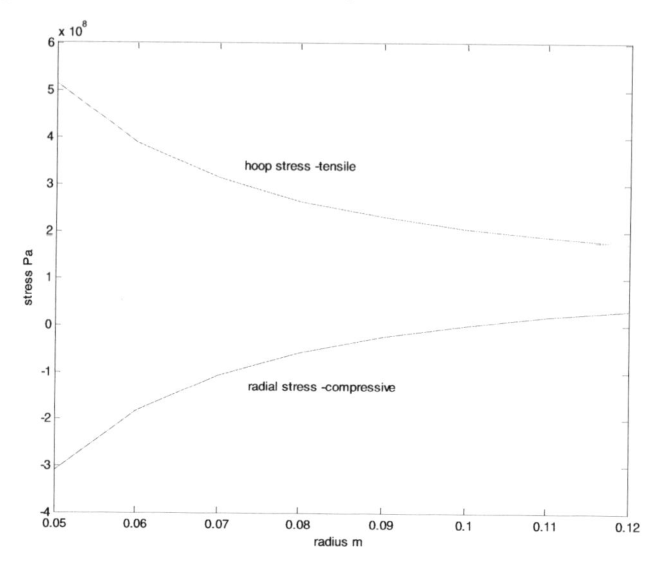

Fig 1.3.2: Hoop stress in monobloc barrel of increased thickness of Example 1.3.1 B

1.4: Built Up Gun Barrels

Built Up or Shrink Fit Gun Barrels

This is a form of pre-stressing in which one or more barrels are built up on each other. The outer radius of the inner cylinder is larger than the inner radius of the outer cylinder. The assembly is done after expansion by heating of the outer cylinder. On cooling a contact pressure is produced, due to the contractive pressure of the outer cylinder, on the inner tube and the expansive pressure exerted by the inner tube, on the outer.

Given that:

- The external radius of the inner cylinder is larger than the inner radius of the outer by an amount δ.
- The pressure produced at the contact surface of the two tubes after assembly of the outer tube on the inner tube is p.

The magnitude of p is found from the condition that:

Increase in inner radius of outer tube + decrease in outer radius of inner tube = δ. Hence from Equations [1.2.16] and [1.2.17]

$$\frac{bp}{E}\left(\frac{b^2+c^2}{c^2-b^2}+\mu\right)+\frac{bp}{E}\left(\frac{a^2+b^2}{b^2-a^2}-\mu\right)=\delta$$

From which:

$$p=\frac{E\delta}{b}\frac{\left(b^2-a^2\right)\left(c^2-b^2\right)}{2b^2\left(c^2-a^2\right)} \quad \text{[1.4.1]}$$

Equations [1.2.12] and [1.2.13] give the stresses in the outer cylinder and Equations [1.2.14] and [1.2.15] give the stresses in the inner cylinder. The stresses considered in design are the stresses at the inner surface of the outer cylinder as the maximum and hence the critical hoop stress is attained here.

$\sigma_t = \dfrac{p\left(b^2+c^2\right)}{c^2-a^2}$ and $\sigma_r = -p$

Example 1.4.1

Compute the hoop stresses in a built up gun barrel with an inner tube of inner radius 4 inches, outer radius 6 inches and an outer tube of inner radius 6 inches and outer radius 8 inches, subject to a maximum gas pressure of 463.2 M Pa, on firing. Displacement at the interface is 0.127mm. *E* for steel used is 206.85 G Pa.

Solution: Example 1.4.1

- The interface pressure is first found using equation [1.4.1]

$$p = \frac{E\delta}{b}\frac{(b^2 - a^2)(c^2 - b^2)}{2b^2(c^2 - a^2)}$$

- Secondly the stresses due to shrinkage are computed using Equations [1.2.14] and [1.2.15] for the inner cylinder, substituting the pressure at the interface p for the external pressure p_0. i.e.

$$\sigma_r = -\frac{p_o b^2}{b^2 - a^2}\left(1 - \frac{a^2}{r^2}\right)$$

And:

$$\sigma_t = -\frac{p_o b^2}{b^2 - a^2}\left(1 + \frac{a^2}{r^2}\right)$$

- For the outer cylinder Equation [1.2.12] is applied for the radial stress, substituting the pressure at the interface p in place of the internal pressure p_i.

$$\sigma_r = \frac{a^2 p_i}{b^2 - a^2}\left(1 - \frac{b^2}{r^2}\right)$$

- Again for the outer cylinder, Equation [1.2.13] is applied for the hoop stress, with substitution for pressure as above:

$$\sigma_t = \frac{a^2 p_i}{b^2 - a^2}\left(1 + \frac{b^2}{r^2}\right)$$

- Next the stresses due to the propellant gas pressure are computed again using Equations [1.2.12] and [1.2.13], in their original forms, using the given value of the maximum gas pressure.

- Finally the resultant stresses are computed by adding the stresses due to shrinkage to the stresses due to maximum propellant gas pressure.

Computer Programme 1.4.1

```
% programme to compute hoop stresses in built up barrel of example
1.4.1
a=.1016 % inner radius; b=.1524 % outer radius; c=.2032 % interface
radius
r=(.1016:.01:.2032);r1=(.1016:.01:.1524);r2=(.1524:.01:.2032)
p_i=463.2.*10^6 % gas pressure; d=.127./1000 %displacement
p=206.85.*10.^9.*d.*(c.^2-b.^2).*(b.^2-a.^2)./(b.*2.*b.^2.*(c.^2-a.^2))%
interface pressure
sigmaro=a.^2.*p./(b.^2-a.^2).*(1-b.^2./r2.^2)% radial stress outer
sigmato=a.^2.*p./(b.^2-a.^2).*(1+b.^2./r2.^2)% hoop stress outer
sigmari=-p.*b.^2./(b.^2-a.^2).*(1-a.^2./r1.^2)%radial stress inner
sigmati=-p.*b.^2./(b.^2-a.^2).*(1+a.^2./r1.^2)%hoop stress
sigmarpi=a.^2.*p_i./(c.^2-a.^2).*(1-c.^2./r.^2)
sigmatp_i1=a.^2.*p_i./(c.^2-a.^2).*(1+c.^2./r1.^2)
sigmatp_i2=a.^2.*p_i./(c.^2-a.^2).*(1+c.^2./r2.^2)
sigmat_res1=sigmatp_i1+sigmati
sigmat_res2=sigmatp_i2+sigmato
plot(r1,sigmatp_i1,r2,sigmatp_i2,r1,sigmat_res1,r1,sigmati,r2,sigmat_res
2,r2,sigmato)
```

Results: Example 1.4.1

Hoop stresses across the barrel wall are tabulated below:

Interface pressure = 27.931 MPa			
Radius m	Hoop stress due to shrink fitting Pa	Hoop stress due to gas pressure Pa*10^8	Resultant hoop stress Pa*10^8
0.1016	-1.0055 *10^8	7.7200	6.7145
0.1116	-0.9195*10^8	6.6628	5.7433
0.1216	-0.8537*10^8	5.8555	5.0018
0.1316	-0.8024*10^8	5.2251	4.4227
0.1416	-0.7616*10^8	4.7236	3.9620
0.1516	0.7286*10^8,4.4690*10^7	4.3179,4.2889	3.589,4.7358
0.1616	4.2023*10^7	3.9613	4.3815
0.1716	3.9806*10^7	3.6890	4.0870
0.1816	3.7944 *10^7	3.4602	3.8397
0.1916	3.6365*10^7	3.2662	3.6298
0.2016	3.5013*10^7	3.1002	3.4504 4

Conclusions

The following conclusions are drawn from Fig 1.4.1:

- Tensile hoop stress results in the outer tube and compressive hoop stress results in the inner tube as a result of shrink fitting.
- As a consequence of shrink fitting the inner tube is pre-stressed compressively.
- The peak stress on the tube during firing is considerably reduced, as compared to monobloc tubes or comparable dimensions, in this case from 350 M Pa to about 250 M Pa.
- On loading, the maximum stress is felt at the inner surface of the outer tube. In other words, if failure occurs, it will initiate at this surface.

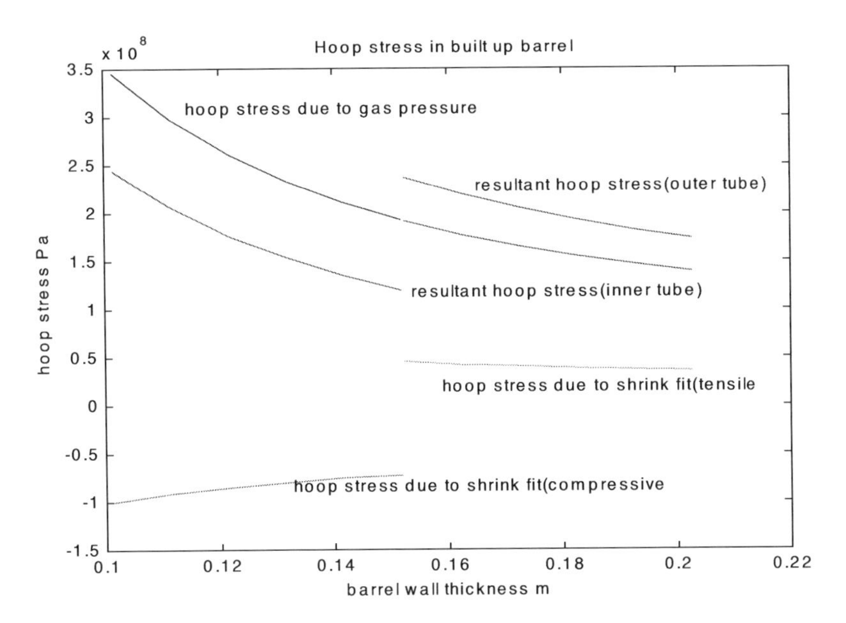

Fig 1.4.1: Hoop stresses in a built up barrel before and during firing

1.5: Autofrettage

Pre-stressing by Autofrettage

Autofrettage is a method of pre-stressing high-pressure pressure vessels such as gun barrels. In this process the interior of the barrel is subject to a high hydraulic or swage pressure. As the pressure inside the gun barrel is gradually increased, a point is reached when the material at the inner surface begins to yield. Yielding is dependant on the pressure reaching a critical value which causes stress in excess of the yield point of the material. With further increase in pressure, the plastic deformation penetrates deeper & deeper into the wall of the cylinder, until some portion of the wall is brought into the plastic condition. If the increase in pressure is unabated, the entire wall becomes plastic. On the other hand, if the pressure is held constant at some intermediate value, it results in a cylindrical region of material, which has become plastic surrounded by a cylindrical region of material, which is still elastic. At the radius of the interface, where yield has just taken place, a radial pressure acts between the two regions, as the elastic region compresses the plastic region, which in turn tends to expand the elastic zone. The situation is similar generally, but only so, to that of a built up barrel.

a: inner radius of tube
c: outer radius of tube
p_{yp}: pressure at yield
p_{af}: autofrettage pressure
b: interface radius
$\tau_{\max}$: maximum shear stress at yield

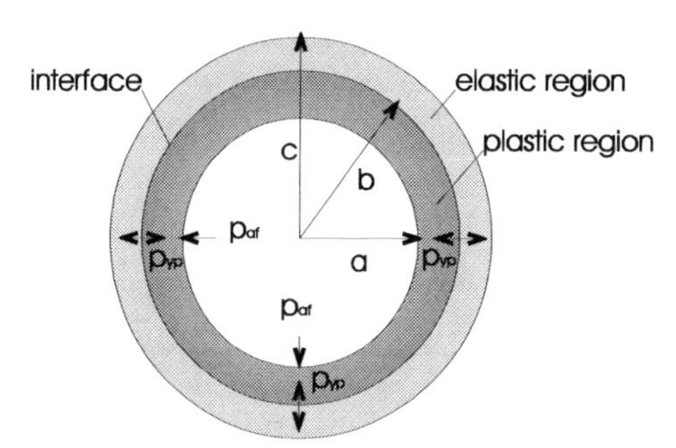

Fig 1.5.1: Regions of a gun barrel after autofrettage

Pressure at the Interface

The magnitude of the pressure p_{yp} at the interface is found from consideration of the elastic outer region of the wall.

From the Tresca theory, the maximum shearing stress at the outer surface of the plastic portion is:

$$\tau_{\max} = \frac{\sigma_t - \sigma_r}{2}$$

Hence:

$$\tau_{\max} = \frac{p_{yp} c^2}{c^2 - b^2}$$

Or:

$$p_{yp} = \frac{\tau_{max}\left(c^2 - b^2\right)}{c^2} \quad \text{[1.5.1]}$$

Stresses due to Deformation

Stresses in the Elastic Region

Knowing the pressure corresponding to the yield point p_{yp} the stresses in the elastic region can be calculated using Equations [1.2.12] and [1.2.13], suitably modified, for a cylinder under internal pressure only, i.e. by using radius b and pressure at yield p_{yp} instead of radius a and pressure at inner surface p_i respectively.

$$\sigma_r = \frac{b^2 p_{yp}}{c^2 - b^2}\left(1 - \frac{c^2}{r^2}\right) \quad \text{[1.5.2]}$$

And:

$$\sigma_t = \frac{b^2 p_{yp}}{c^2 - b^2}\left(1 + \frac{c^2}{r^2}\right) \quad \text{[1.5.3]}$$

Stresses in the Plastic Region

For every point in the plastic region, the following equation holds good:

$$\tau_{max} = \frac{\sigma_t - \sigma_r}{2}$$

From Equation [1.2.1], we have:

$$\sigma_t - \sigma_r - r\frac{d\sigma_r}{dr} = 0$$

Hence:

$$\frac{d\sigma_r}{dr} = \frac{2\tau_{max}}{r}$$

Integrating:

$$\sigma_r = 2\tau_{max} \ln r + C$$

The constant of integration C is found from the condition:

For $r = b$, $\sigma_r = -p_{yp}$:

$-p_{yp} = 2\tau_{max} \ln b + C$ or $C = -p_{yp} - 2\tau_{max} \ln b$

Hence:

$$\sigma_r = 2\tau_{max} \ln\frac{r}{b} - \frac{\tau_{yp}\left(c^2 - b^2\right)}{c^2} \qquad [1.5.4]$$

And:

$$\sigma_t = 2\tau_{max} + \sigma_r \qquad [1.5.5]$$

Autofrettage Pressure

If p_{af} be the pressure applied to cause yield up to a radius b, then the radial stress at the inner surface of radius a, $\sigma_r = -p_{af}$.

Hence from Equation [1.5.4]:

$$\sigma_r = 2\tau_{max} \ln\frac{a}{b} - \frac{\tau_{max}\left(c^2 - b^2\right)}{c^2}$$

The pressure of autofrettage is given by:

$$p_{af} = -2\tau_{max} \ln\frac{a}{b} + \frac{\tau_{max}\left(c^2 - b^2\right)}{c^2} \qquad [1.5.6]$$

Residual Stresses after Unloading of Autofrettage Pressure

After partial yield of the wall, if the autofrettage pressure is removed, a residual stress remain in the wall of the barrel as the inner portion has undergone plastic deformation and now undergoes a pressure from the elastic outer portion. The residual stress distribution after unloading is calculated by unloading the autofrettage pressure by subtracting the stresses using the equations for a cylinder subject to internal pressure, in this case the autofrettage pressure p_{af}.

Residual Stresses in the Elastic Region

Residual stresses in the elastic region are computed by subtracting the stresses due to autofrettage from the stresses due to deformation and autofrettage pressure using Equations [1.2.12] and [1.2.13]. In the elastic region the radius r varies from b to c. The subscript e here denotes the elastic region.

$$\sigma_{r(res)e} = \frac{b^2 p_{yp}}{c^2 - b^2}\left(1 - \frac{c^2}{r^2}\right) - \frac{a^2 p_{af}}{c^2 - a^2}\left(1 - \frac{c^2}{r^2}\right) \qquad [1.5.7]$$

And:

$$\sigma_{t(res)e} = \frac{b^2 p_{tp}}{c^2 - b^2}\left(1 + \frac{c^2}{r^2}\right) - \frac{a^2 p_{af}}{c^2 - a^2}\left(1 + \frac{c^2}{r^2}\right) \qquad [1.5.8]$$

Residual Stresses in the Plastic Region

In the plastic region, the radius r varies from a to b. The subscript p denotes the plastic region.

$$\sigma_{r(res)p} = 2\tau_{\max} \ln\frac{r}{b} - \frac{\tau_{\max}\left(c^2 - b^2\right)}{c^2} - \frac{a^2 p_{af}}{c^2 - a^2}\left(1 - \frac{c^2}{r^2}\right) \qquad [1.5.9]$$

And:

$$\sigma_{t(res)p} = 2\tau_{max} + \sigma_r - \frac{a^2 p_{af}}{c^2 - a^2}\left(1 + \frac{c^2}{r^2}\right) \qquad [1.5.10]$$

Working Stresses

The autofrettaged barrel is subject to internal pressure of the propellant gases during firing. The stresses now experienced in the gun are the sum of the residual stresses and the stresses due to propellant gas pressure. The stresses due to gas pressure are computed, using the condition that the barrel is subject to internal pressure alone, from Equations [1.2.12] and [1.2.13], p_i here being the maximum propellant gas pressure.

Working Stresses in the Elastic Region

The radial stresses in the elastic region are given as under:

$$\sigma_{r(work)e} = \sigma_{r(res)e} + \frac{a^2 p_i}{c^2 - a^2}\left(1 - \frac{c^2}{r^2}\right) \qquad [1.5.11]$$

The hoop stresses are given by:

$$\sigma_{t(work)e} = \sigma_{t(res)e} + \frac{a^2 p_i}{c^2 - a^2}\left(1 + \frac{c^2}{r^2}\right) \qquad [1.5.12]$$

Working Stresses in the Plastic Region

Radial stresses in the plastic region are as follows:

$$\sigma_{r(work)p} = \sigma_{r(res)p} + \frac{a^2 p_i}{c^2 - a^2}\left(1 - \frac{c^2}{r^2}\right) \qquad [1.5.13]$$

And the hoop stresses are given by:

$$\sigma_{t(work)p} = \sigma_{t(res)p} + \frac{a^2 p_i}{c^2 - a^2}\left(1 + \frac{c^2}{r^2}\right) \qquad [1.5.14]$$

Example 1.5.1

Compute and plot the hoop stresses due to deformation, the residual stresses and the working stresses, against wall thickness, in an autofrettaged gun barrel of the following dimensions; inner radius 4 inches, outer radius 8 inches, interface radius 6 inches, given the maximum shear stress of the steel used is 425 MPa, and the maximum propellant gas pressure is 240 MPa.

Solution: Example 1.5.1

- The pressure at the interface is computed using Equation [1.5.1].
- Knowing this pressure the autofrettage stresses can be calculated with the help of Equations [1.5.2], [1.5.3], [1.5.4] and [1.5.5].
- Next the autofrettage pressure is calculated using equation [1.5.6].
- The residual stresses after unloading the autofrettage pressure are now calculated by equations [1.5.9] and [1.5.10] for the plastic region and [1.5.7] and [1.5.8] for the elastic region.
- Finally the working stresses are computed by summing the residual and the propellant gas pressure stresses with the help of equations [1.5.11] and [1.5.12] for the elastic region and [1.5.13] and [1.5.14 for the plastic region.

Computer Programme 1.5.1

```
% programme to compute hoop stresses after autofrettage
a=0.1016 % inner radius m
b=0.1524 % interface radius m
c=0.2032 % outer radius m
r=(.1016:.01:.2032) % radius array
r1=(.1016:.01:.1524) % radius array plastic region
```

```
r2=(.1524:.01:.2032) % radius array elastic region
tow_yp=425*10^6 % max shear stress Pa
pw=240*10^6 % max gas pressure Pa
% pressure at interface:
pb=tow_yp.*(c.^2-b.^2)/c.^2
% elastic region: af  stress in outer tube
sigmato=b.^2.*pb./(c.^2-b.^2).*(1+c.^2./r2.^2)
% plastic region af stress in inner tube
sigmari=2.*tow_yp.*log(r1./b)-tow_yp.*(c.^2-b.^2)./c.^2
sigmati=2.*tow_yp+sigmari
%stress due to af pressure
%Pressure causing plastic def upto radius b
paf=-2.*tow_yp.*log(a./b)+tow_yp.*(c.^2-b.^2)/c.^2
sigmat_inner=a^2.*paf/(c.^2-a.^2).*(1+c.^2./r1.^2)
sigmat_outer=a^2.*paf/(c.^2-a.^2).*(1+c.^2./r2.^2)
% residual stresses
sigmat_resinner=sigmati-sigmat_inner
sigmat_resouter=sigmato-sigmat_outer
% working stress
sigmat_workinner=a.^2.*pw./(c.^2-a.^2).*(1+c.^2./r1.^2)
sigmat_workouter=a.^2.*pw./(c.^2-a.^2).*(1+c.^2./r2.^2)
% final working stress
sigma_finalinner=sigmat_workinner+sigmat_resinner
sigma_finalouter=sigmat_workouter+sigmat_resouter
plot(r2,sigmato,r1,sigmati,r1,sigmat_inner,r2,sigmat_outer,r1,sigmat_res
inner,r2,sigmat_resouter,r1,sigmat_workinner,r2,sigmat_workouter,r1,si
gma_finalinner,r2,sigma_finalouter)
```

Tabulated results of Computer Programme 1.5.1 are given below:

Pressure at the interface = $1.8594*10^8$ Pa				
Autofrettage pressure = $5.3058*10^8$ Pa				
Rad-ius	**Hoop stress due to yielding Pa*10^8**	**Residual hoop stress after unloading of af pressure Pa*10^8**	**Hoop stress due to propellant gas pressure Pa*10^8**	**Final resultant hoop stress Pa*10^8**
0.1016	8.8430	-5.6489	4.0000	-1.6489
0.1116	7.6320	-3.6399	3.4522	-0.1877
0.1216	6.7073	-1.9857	3.0339	1.0482
0.1316	5.9853	-0.5919	2.7073	2.1154
0.1416	5.4107	0.6051	2.4474	3.0526
0.1516	4.9461,4.9128	1.6498,1.7278	2.2373,2.2222	3.8871,3.9500
0.1616	4.5375	1.5958	2.0525	3.6483
0.1716	4.2256	1.4861	1.9114	3.3975
0.1816	3.9636	1.3940	1.7929	3.1868
0.1916	3.7413	1.3158	1.6923	3.0082
0.2016	3.5512	1.2490	1.6063	2.8553

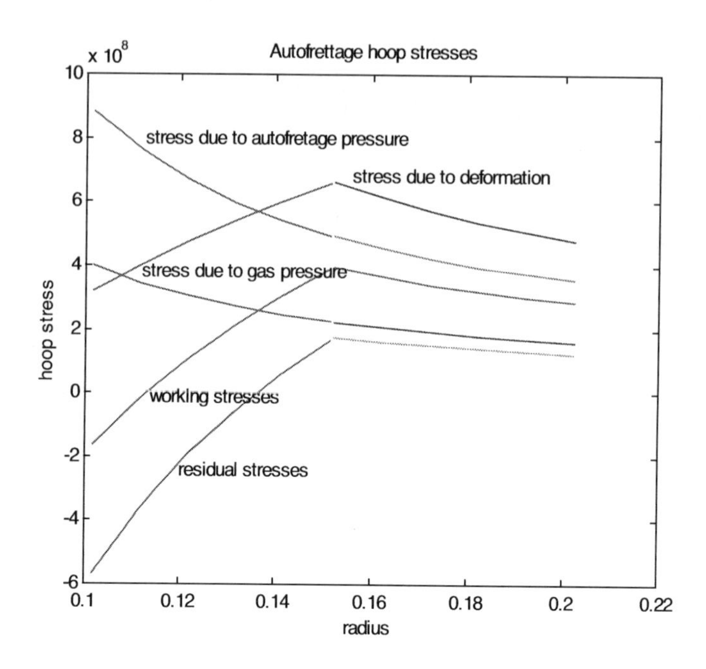

Fig 1.5.2: Graphical representation of stresses in autofrettaged gun barrel of Example 1.5.1

Conclusions

- The residual stress in an autofrettaged barrel is compressive for the most part in the plastic region; however the point of inflexion occurs before the interface.
- The critical or the radius of maximum stress shifts to the inner surface of the elastic region.
- The material of the barrel is more uniformly stressed throughout its thickness than in the case of monobloc construction.
- The optimum stress condition may be obtained by careful selection of suitable values of the interface radius by varying the autofrettage pressure accordingly.

1.6: Rifling

Rifling

Rifling is the system of spiral grooves and the raised portion between the grooves called the lands on the bore of the barrel. The rifling in conjunction with the driving band, of the projectile, imparts the necessary spin to stabilize the projectile along its longer axis with respect to the trajectory during its flight. There exists an optimum spin rate below which the projectile is under-stabilized and above which the projectile is over stabilized. Under-stabilization implies that the projectile will tumble in flight. Over-stabilization implies that the projectile longer axis maintains a rigid attitude through out the flight resulting in a broad side on, rather than a nose first impact, at the target.

Often, the centrifugal force arising out of the spin imparted to the projectile is used to perform certain important functions concerning arming of the fuze in flight. This is the reason why some projectiles even though primarily fin stabilized, are imparted a limited spin rate in the barrel.

Definitions Associated with Rifling

- Angle of rifling. The angle between the groove or the tangent to the curve of the groove at any point on the bore surface with the line parallel to the bore axis and passing through the point in question.

- Pitch of rifling. The distance, which may be referred to in units of calibres, in which the groove makes one complete revolution.

- Twist of rifling. This is the number of turns that the groove makes per calibre length and is the reciprocal of the pitch of rifling.

Types of Rifling

Rifling is defined with respect to the angle the tangent to the spiral curve of the groove, makes with the axis of the bore. Rifling may be of uniform twist, when the slope of the tangent to the rifling curve is constant, or varying when the slope of the tangent to the rifling curve changes, or a combination of both uniform and varying twist. Varying twist may be parabolic, cubic or semi cubic depending on the mathematical expression used to define the rifling curve.

If a barrel is slit along it longer axis and the cylinder unrolled flat, as in Fig 1.6.1 the system of lands and grooves will lie on the plane of the sheet, enabling easy visualization of the angle of twist of the rifling.

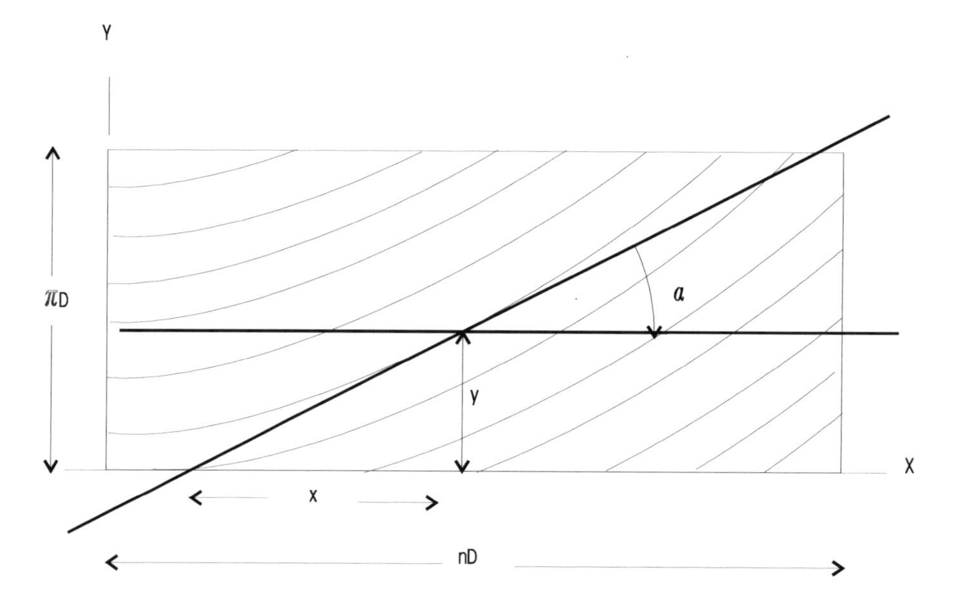

Fig 1.6.1: Angle of twist of rifling

Linear Velocity and Angular Velocity of the Projectile

The rate of spin imparted to the projectile is a function of the linear velocity of the projectile at exit from the barrel as also the angle of the spiral of the rifling.

The linear velocity of rotation of the projectile is given by:

$$v_r = v_{(e)} Tan\alpha = \frac{\pi}{n} v_{(e)}$$

$$\because Tan\alpha = \frac{\pi D}{nD}$$

$v_{(e)}$: muzzle velocity of the projectile
α: angle of rifling
n: pitch of rifling
D: calibre

If a particle in circular motion describes a small angle $d\theta$ in a time interval dt, the angular velocity of the particle is given by:

$$\omega = \frac{d\theta}{dt}$$

The linear rotational velocity is given by the length of the arc described divided by the time interval i.e.

$$v_r = r\frac{d\theta}{dt}$$

Hence the angular velocity is given by:

$\omega = \frac{v_r}{r}$ radians per second

Or:

$\omega = \frac{2\pi}{nD} v_{(e)}$ radians per second

Hence the spin rate is given by:

$N = \frac{1}{nD} v_{(e)}$ revolutions per second.. [1.6.1]

Example 1.6.1

Calculate the spin rate of a 130 mm projectile with a muzzle velocity of 930 m/s if the pitch of rifling is 30 calibres.

Solution: Example 1.6.1

Using Equation [1.6.1]:

$$N = \frac{1}{30.130} 930.1000 = 238.46 \text{ RPS}$$

Or:

$N = 14307$ RPM

Equation and Slope of the Rifling Curve

Uniform Twist

In the case of uniform twist, the angle of twist of rifling is constant so that the tangent of the angle of rifling:

$\frac{dy}{dx} = \text{constant} = \frac{\pi}{n}$ .... [1.6.2]

n: distance along the barrel axis in which a groove makes one complete turning multiples of calibre.

Integrating Equation [1.6.2]:

$$y = \frac{\pi}{n} x + C \quad \text{.......... [1.6.3]}$$

C is the intercept on the Y-axis and denotes the point of origin of the groove on the inner circumference of the barrel.

Parabolic or Uniformly Increasing Twist

In this case the slope of the tangent to the groove increases uniformly or

$$\frac{d^2y}{dx^2} = C_1 = \text{a constant.}$$

Integrating for the first time:

$$\frac{dy}{dx} = C_1 x + C_2$$

Integrating again:

$$y = C_1 \frac{x^2}{2} + C_2 x + C_3$$

C_3 corresponds to the constant C of Equation [1.6.3], in the case of uniform twist.

If the boundary conditions are:

Breech end:

$$\frac{dy}{dx} = \frac{\pi}{n_1} \text{ and } x = 0$$

Muzzle end:

$$\frac{dy}{dx} = \frac{\pi}{n_2} \text{ and } x = l$$

Then:

$$C_2 = \frac{\pi}{n_1}$$

$$\frac{dy}{dx} = \frac{\pi}{n_2} = C_1 l + \frac{\pi}{n_1}$$

Or:

$$C_1 = \frac{\pi}{l}\left(\frac{1}{n_2} - \frac{1}{n_1}\right)$$

l: length of the rifled portion of the barrel.

The equation for parabolic rifling now becomes:

$$y = \frac{\pi}{l}\left(\frac{1}{n_2} - \frac{1}{n_1}\right)\frac{x^2}{2} + \frac{\pi}{n_1}x + C_3 \qquad [1.6.4]$$

The slope of the rifling curve at any point is given by:

$$\frac{dy}{dx} = \frac{\pi}{l}\left(\frac{1}{n_2} - \frac{1}{n_1}\right)x + \frac{\pi}{n_1} \qquad [1.6.5]$$

Cubic Twist

This is a form of twist in which the angle of rifling increases progressively. The curve of the rifling groove is defined by the expression:

$$y = C_1x^3 + C_2x^2 + C_3x + C_4$$

For this curve, the greatest exponent of x is 3.

If the point of inflexion of this curve is at the muzzle, then at this point:

$$\frac{d^2y}{dx^2} = 0$$

Differentiating the general equation for the cubic curve:

$$\frac{dy}{dx} = 3C_1x^2 + 2C_2x + C_3$$

If the pitch of rifling at the breech end is $\frac{\pi}{n_1}$.

Then at the breech end:

$$\frac{dy}{dx} = \frac{\pi}{n_1} \text{ as } x = 0$$

If the angle of twist at the muzzle end is $\frac{\pi}{n_2}$ and l the length of rifling,

then at the muzzle end:

$$\frac{d^2y}{dx^2} = \frac{\pi}{n_2} = 3C_1l^2 + 2C_2l + \frac{\pi}{n_1}$$

Or:

$$\frac{\pi}{n_2} - \frac{\pi}{n_1} = 3C_1 l^2 + 2C_2 l .$$

From the point of inflexion at the muzzle:

$$\frac{dy}{dx} = 6C_1 l + 2C_2 \text{ or } 3C_1 l = -C_2$$

Hence:

$$C_1 = -\frac{\frac{\pi}{n_2} - \frac{\pi}{n_1}}{3l^2} \quad and \quad C_2 = \frac{\frac{\pi}{n_2} - \frac{\pi}{n_1}}{l} .$$

The equation for the rifling curve now becomes:

$$y = \left(\frac{\pi}{n_2} - \frac{\pi}{n_1}\right)\left(\frac{x^2 l}{} - \frac{3x^3}{3l^2}\right) + \frac{\pi}{n_1}x \quad \text{.................... [1.6.6]}$$

The slope at any point is given by:

$$\frac{dy}{dx} = \left(\frac{\pi}{n_2} - \frac{\pi}{n_1}\right)\left(\frac{2xl}{l} - \frac{2x^3}{3l^2}\right) + \frac{\pi}{n_1} \quad \text{.................... [1.6.7]}$$

Semi Cubic Twist

The curve for this kind of rifling is defined by the exponent $\frac{3}{2}$ of the displacement x along the bore in the axial direction:

The general equation for this curve is $y = C_1 + C_2(x + C_3)^{\frac{3}{2}}$

And:

$$\frac{dy}{dx} = \frac{3}{2}C_2(x + C_3)^{\frac{1}{2}}$$

The constants are determined from the boundary conditions and from the initial value $y = 0$.

$$C_1 = -\frac{2}{3}\frac{\frac{\pi}{n_2}\left(\frac{\frac{\pi}{n_1}}{\frac{\pi}{n_2}}\right)^3 l}{\left(1 - \frac{\left(\frac{\pi}{n_1}\right)^2}{\left(\frac{\pi}{n_2}\right)^2}\right)} \; ; \; C_2 = \frac{2}{3}\frac{\frac{\pi}{n_2}}{l^{\frac{1}{2}}}\left(1 - \frac{\left(\frac{\pi}{n_1}\right)^2}{\left(\frac{\pi}{n_2}\right)^2}\right)^{\frac{1}{2}} \; ; \; C_3 = \frac{l\frac{\left(\frac{\pi}{n_1}\right)^2}{\left(\frac{\pi}{n_2}\right)^2}}{1 - \frac{\left(\frac{\pi}{n_1}\right)^2}{\left(\frac{\pi}{n_2}\right)^2}}$$

Knowing the values of the constants C_1, C_2 and C_3, the equations for the slope of the rifling curve and the curve itself may be written.

Example 1.6.2

Plot the rifling curves for rifling with uniform twist and with parabolic twist given the following data:

Length of rifled portion of barrel: 5.86 m

For parabolic twist:

Pitch of rifling at breech end: 30 calibres
Pitch of rifling at muzzle end: 20 calibres

For uniform twist:

Pitch of rifling: 20 calibres.

Solution: Example 1.6.2

- An array with suitable increments is constructed with values varying from 0 at the breech end to the length of the rifled portion at the muzzle end.
- The curve for rifling with uniform twist is computed using Equation [1.6.3].
- The curve for rifling with parabolic twist is computed using Equation [1.6.4]

Computer Programme 1.6.1

```
% programme to plot rifling curves
l=5.86 % length of rifled portion
n1=30 % pitch of rifling at breech
n2=20 % pitch of rifling at muzzle
x=linspace(0,l,100) % axial displacement array
y1=pi./n2.*x % curve for uniform twist
y=pi./l.*((1./n2)-(1./n1)).*(x.^2)./2+pi./n1.*x % curve for parabolic twist
plot(x,y1,x,y)
title('Rifling curves for uniform and parabolic twist')
xlabel('length of rifled portion')
ylabel('circumference')
gtext('uniform twist')
gtext('parabolic twist')
```

Results: Example 1.6.2

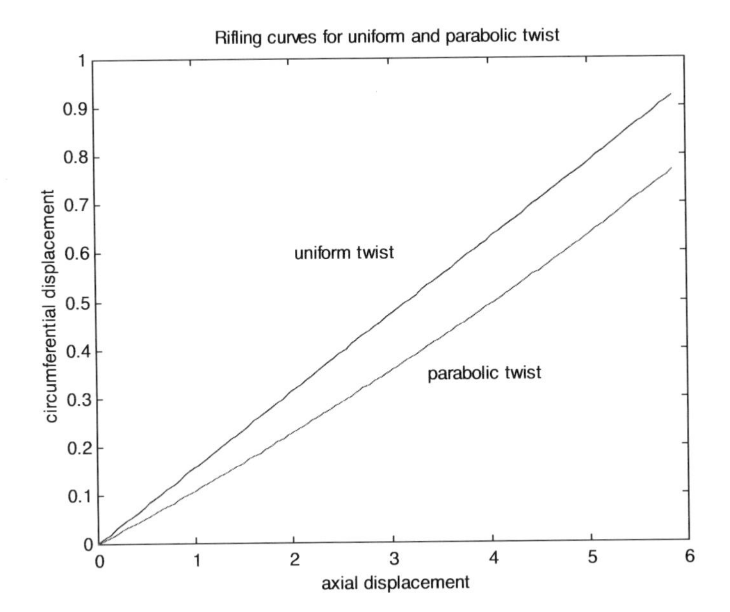

Fig 1.6.2: Rifling curves, parabolic and uniform twist

The slope of both curves at the muzzle is identical. It follows that the spin rate of the projectile will also be equal in both cases. As will be seen in the next section, in the case of parabolic rifling, the force exerted on the projectile at exit is greater than in the case of uniform twist. This results in comparatively greater instability of the projectile. In the case of uniform rifling, the stresses on the lands and driving band of the projectile are greater towards the initial stages, while the destabilizing force on the projectile at the muzzle is lesser.

1.7: Force on the Driving Band

Thrust on the Lands and Force on the Driving Band

The thrust on the lands caused by the propellant gas force acting on the projectile base and the angle of rifling and the consequent reaction force on the driving band of the projectile depend on the twist of rifling for a given gas pressure pattern. Uniform twist of rifling has the major disadvantage of heavy loading on the lands especially around the commencement of rifling. In the case of increasing twist, the thrust on the lands and the reaction on the driving band can be reduced towards the initial stages of engraving. From the accuracy point of view it is desirable to have the minimum forces acting on the projectile at the instant when the guidance of the bore ceases i.e. at projectile exit. In the case of uniform twist of rifling, the force on the driving band at exit is less than in the case of increasing twist of rifling. Hence a combination of the two is often advocated. This entails an initial section of increasing twists followed by a section of uniform twist which continues up to the muzzle.

Equation for Force on the Driving Band

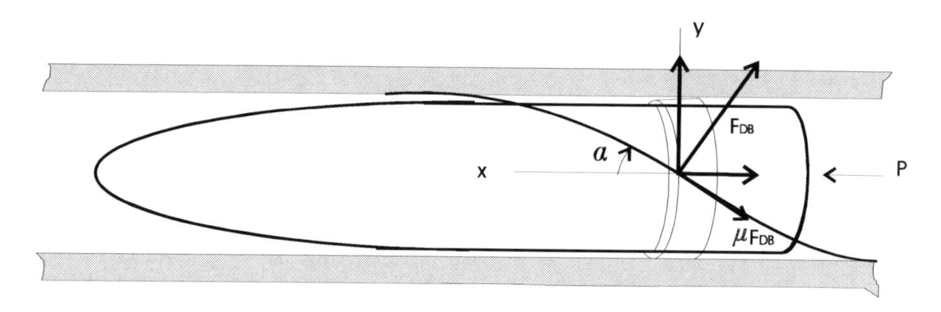

Fig 1.7.1: Forces on the projectile body

P: force due to propellant gas pressure on projectile base.
α: twist angle.
x: direction of shot travel.
y: tangential displacement of a point on the circumference of the projectile.
F_{DB}: force on driving band
μR: frictional force.
D: calibre.
$I = m_p.k^2$: projectiles moment of inertia about its longitudinal axis.
ω: angular velocity.
m_p: mass of projectile.
v_p: instantaneous velocity of projectile
k: radius of gyration of projectile; $k^2 = \frac{I}{m_p}$

Taking the axis of the bore as the X-axis and resolving the forces in this direction, gives:

$$P - F_{DB}\,(Sin \propto + \mu Cos \propto) = m_p \frac{dv_p}{dt}$$

Similarly resolving the forces in the direction of the Y-axis:

$$F_{DB}(Cos \propto - \mu Sin \propto)\frac{D}{2} = I\frac{d\varpi}{dt}$$

Also:

$$\omega = \frac{2}{D}v_p \tan\alpha = \frac{2}{D}v_p \frac{dy}{dx}.$$

$$\frac{d\omega}{dt} = \frac{2}{D}\left[\frac{dv_p}{dt}.\frac{dy}{dx} + v_p \frac{d\left(\frac{dy}{dx}\right)}{dt}\right]$$

$$\frac{d\omega}{dt} = \frac{2}{D}\left[\frac{dv_p}{dt}.\frac{dy}{dx} + v_p \frac{d\left(\frac{dy}{dx}\right)}{dt}\frac{dx}{dx}\right]$$

$$\frac{d\omega}{dt} = \frac{2}{D}\left[\frac{dv_p}{dt}.\frac{dy}{dx} + v_p \frac{dx}{dt}\frac{d\left(\frac{dy}{dx}\right)}{dx}\right]$$

$$\frac{d\omega}{dt} = \frac{2}{D}\left[\frac{dv_p}{dt}.\frac{dy}{dx} + v_p^2 \frac{d^2y}{dx^2}\right]$$

The angle of rifling is generally small so in order to simplify the analysis approximations can be made as under:

$Sin\alpha = 0$, $Cos\alpha = 1$, $\mu F_{DB} < P$; F_{DB} is negligible with respect to P:

$$P = m_p \frac{dv_p}{dt}$$

$$F_{DB} = m_p \{\frac{2k}{D}\}^2 \left[\frac{dv_p}{dt}.\frac{dy}{dx} + v_p^2 \frac{d^2y}{dx^2}\right]$$

$$F_{DB} = \{\frac{2k}{D}\}^2 \left[\frac{dy}{dx}.P + m_p v_p^2 \frac{d^2y}{dx^2}\right] \qquad \text{[1.7.1]}$$

Equation [1.7.1] represents a formula for the force on the driving band irrespective of the nature of the rifling.

Force on the Driving Band: Rifling with Constant Twist

The force on the driving band in the particular case of rifling with uniform twist can be obtained by substituting the known values of slope and the rate of change of slope for rifling with uniform twist in the general Equation [1.7.1], for force on the driving band.

Equation for the rifling curve: $y = Cx = \frac{\pi}{n}x$

$$\frac{dy}{dx} = \frac{\pi}{n}, \quad \frac{d^2y}{dx^2} = 0 \text{ as } n \text{ is a constant.}$$

$$F_{DB} = \left(\frac{2k}{D}\right)^2 P\frac{\pi}{n} \quad \text{.......... [1.7.2]}$$

Since n is a constant, the force on the driving band F_{DB} is proportional only to the gas force P or the gas pressure p.

Force on the Driving Band: Parabolic Twist of Rifling

Similarly as in the case of rifling with uniform twist, the force on the driving band in the case of rifling with parabolic twist can be obtained by substituting the known values of slope and rate of change of slope for rifling with parabolic twist in the general Equation [1.7.1], for force on the driving band.

In this case:

$$\frac{d^2y}{dx^2} = C_1 \ and \ \frac{dy}{dx} = C_1x + C_2$$

$$C_2 = \frac{\pi}{n_1} and \, C_1 = \frac{1}{l}\left[\frac{\pi}{n_2} - \frac{\pi}{n_1}\right]$$

Or:

$$\frac{d^2y}{dx^2} = \frac{1}{l}\left[\frac{\pi}{n_2} - \frac{\pi}{n_1}\right]$$

And:

$$\frac{dy}{dx} = \frac{1}{l}\left[\frac{\pi}{n_2} - \frac{\pi}{n_1}\right]x + \frac{\pi}{n_1}$$

Substituting for $\frac{dy}{dx}$ and $\frac{d^2y}{dx^2}$ in the general equation [1.7.1] for force on the driving band:

$$F_{DB} = \left(\frac{2k}{D}\right)^2 \left[\left[\frac{1}{l}\left[\left(\frac{\pi}{n_2} - \frac{\pi}{n_1}\right)x + \frac{\pi}{n_1}\right]\right]P + m_p v_p^2 \frac{1}{l}\left(\frac{\pi}{n_2} - \frac{\pi}{n_1}\right)\right]$$

$$F_{DB} = \left(\frac{2k}{D}\right)^2 \left[\frac{Px + m_p v_p^2}{l}\left(\frac{\pi}{n_2} - \frac{\pi}{n_1}\right) + P\frac{\pi}{n_1}\right] \quad \text{.. [1.7.3]}$$

Similarly the force on the driving band can be established for other rifling curves with twist defined.

Example 1.7.1

Given the following data, plot the force on the driving band against shot travel. Assume the pressure at the projectile base is the chamber pressure.

Uniform rifling:
Angle of twist: 5°
Increasing twist:
Angle of twist at the breech end: 3°
Angle of twist at muzzle end: 5°
Mass of the high explosive projectile: 6.2 kg
Length of the rifled portion: 2.585 m
For the HE shell $2k/D$ ratio = 0.73

The pressure, corresponding projectile travel and velocities may be obtained from the internal ballistic solution given in the worksheet on the next page. The Heydenriech method was used to generate the projectile travel, pressure and projectile velocity data.

Symbols used in the work sheet:

ve: muzzle velocity
vm: projectile velocity
mc: charge mass
pm: peak pressure
sm: projectile travel at instant of peak pressure
lamda: projectile travel ratio
tm: time up to instant of peak pressure
mp: projectile mass
mg: mass of gun
l: total projectile travel in the barrel
p: instantaneous chamber pressure
vp: projectile velocity at any instant.

The other symbols represent Heydenreich's factors for computation of the pressure displacement curve.

Spreadsheet to compute displacement, pressure, velocity					
ve	680	vm	258.4	l	2.587
mc	1.08	tm	2.48E-03	sigmanetap	0.0875
pm	2.27E+08	mp	6.2		
sm	2.26E-01	mg	402		
lamda	s	psi	phi	p	vp
	0		0	0.00E+00	0.000
0.25	0.057	0.741	0.392	1.68E+08	101.293
0.50	0.113	0.912	0.635	2.07E+08	164.084
0.75	0.170	0.980	0.834	2.22E+08	215.506
1.00	0.226	1.000	1.000	2.27E+08	258.400
1.25	0.283	0.989	1.140	2.24E+08	294.576
1.50	0.340	0.965	1.262	2.19E+08	326.101
1.75	0.396	0.932	1.366	2.12E+08	352.974
2.00	0.453	0.898	1.468	2.04E+08	379.331
2.50	0.566	0.823	1.632	1.87E+08	421.709
3.00	0.679	0.747	1.763	1.70E+08	455.559
3.50	0.792	0.675	1.875	1.53E+08	484.500
4.00	0.905	0.604	1.983	1.37E+08	512.407
4.50	1.019	0.546	2.068	1.24E+08	534.371
5.00	1.132	0.495	2.140	1.12E+08	552.976
6.00	1.358	0.403	2.269	9.15E+07	586.310
7.00	1.585	0.338	2.363	7.67E+07	610.599
8.00	1.811	0.284	2.445	6.44E+07	631.788
9.00	2.037	0.248	2.509	5.63E+07	648.326
10.00	2.264	0.220	2.566	4.99E+07	663.054
11.00	2.490	0.199	2.615	4.52E+07	675.716
11.43	2.587	0.191	2.633	4.34E+07	680.367

Fig 1.7.2: Spreadsheet to compute internal ballistics of Example 1.7.1

Solution: Example 1.7.1

The example is solved as follows:

- Shot travel, pressure and velocity are read as variables, direct from the spreadsheet using the wk1read command of Matlab.
- Value of n for uniform twist is calculated.
- Values of n_1 and n_2 for parabolic twist are calculated.
- Force on the driving band is computed using Equation [1.7.2] for uniform twist.
- Force on the driving band is computed using Equation [1.7.3] for parabolic twist.
- Force on the driving band in the case of uniform twist and in the case of parabolic twist are plotted against shot travel

Computer Programme 1.7.1

```
%programme to compute force on the driving band
thetac=5%constant angle of twist degrees
n2=pi./tan(thetac/180.*pi)
theta0=3 %angle of twist at breech degrees
n1=pi./tan(theta0/180.*pi)
thetae=5 %angle of twist at muzzle degrees
mp=6.2 % mass of shell kg
l=2.585
twok_by_D=.73 % HE shell
rng='e4..e25'
rng1='b4..b25'
rng2='f4..f25'
p=wk1read('c:\sol\76mm',3,4,rng) % sol.wk1, 1-2-3 filename
P=p.*.076.^2.*pi./4
s=wk1read('c:\sol\76mm',3,1,rng1)
v=wk1read('c:\sol\76mm',3,5,rng2)
P=p.*.076.^2.*pi./4
Fdb1=twok_by_D^2*P.*pi./n2
Fdb2=twok_by_D.^2.*((((P.*s+mp.*v.^2)./l).*(pi./n2-pi./n1))+P.*pi./n1)
plot(s,Fdb,s,Fdb2)
```

```
title('Force on driving band versus shot travel')
xlabel('shot travel')
ylabel('Force on driving band')
gtext('uniform twist');gtext('parabolic twist')
```

Results: Example 1.7.1

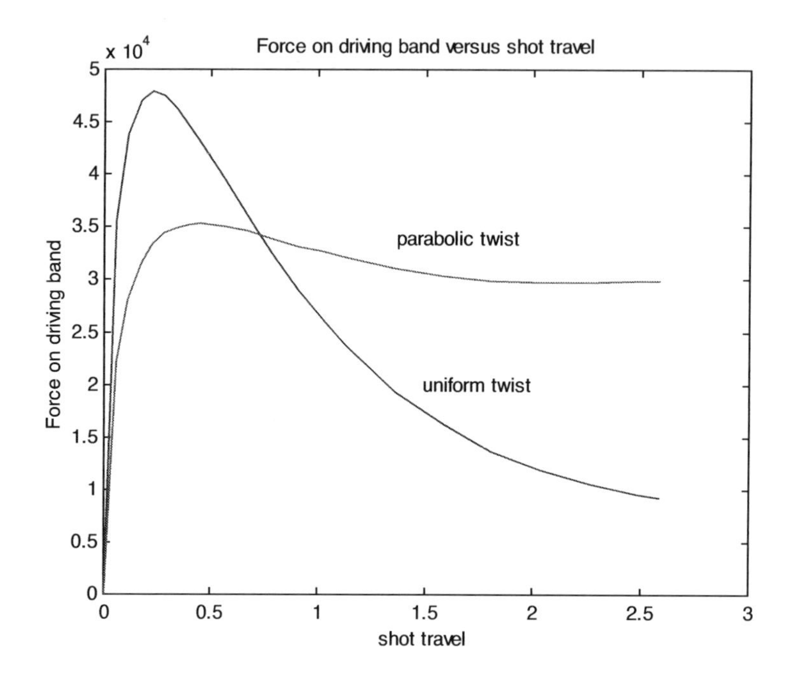

Fig 1.7.3: Force on driving band versus shot travel curve

The curves for uniform twist and parabolic twist represented above clearly illustrate the advantages of lower initial force on the driving band in the case of parabolic twist of rifling, but larger force on the driving band at projectile exit in the case of parabolic twist of rifling as compared to uniform twist of rifling.

1.8: Droop of Gun Barrels

Droop of Gun Barrels

Droop of a gun barrel is the deflection of the axis of a barrel, due to the weight of the barrel itself and that of barrel attachments, if any, from its undeformed or intended axis. Droop may be measured vertically in length units at points along the length of the barrel or as an angular deflection between the intended axis and the line joining the centre of the muzzle to the centre of the barrel at the breech end, measured at the breech end.

Barrel Bend

Barrel bend is dissimilar from droop. Barrel bend is the temporary deformation of the barrel due to ambient temperature differentials at various points on its surfaces. Barrel bend varies according to instantaneous temperature variations. With high accuracy guns it is taken into account by muzzle references incorporated into the fire control system. Muzzle references measure the instantaneous variation between the bore axis at the muzzle and the intended axis at the time of aiming with the help of optical and electro optic devices. The error induced by the barrel bend is corrected for in the central ballistic computer. Barrel bend is also minimized by the use of insulating thermal jackets on the barrel.

Estimation of Droop of a Gun Barrel

In order to estimate the droop it is convenient if the origin of the coordinate system is taken at the breech end, of the barrel with the X-axis as the intended axis of the bore. The deflection y at the muzzle and at points along the barrel distant from the origin can be estimated, as contained further in this section.

Estimation of droop is possible by resort to the theory of pure bending of a beam. Pure bending means the absence of vertical shearing forces acting on the beam. In the case of pure bending of a beam, we have the equation:

$$\frac{E}{r} = \frac{M}{I}$$

Or:

$$\frac{1}{r} = \frac{M}{EI} \qquad \text{[1.8.1]}$$

E: Young's Modulus
I: moment of inertia of the cross section of the beam about the neutral axis
r: radius of curvature of the deflection curve of the beam
x,y: coordinates of any point under consideration on the beam
M: bending moment at the cross section containing the point under consideration

For a uniform beam, the bending moment *M* is constant along its entire length and the radius of curvature is also constant. In the case of a gun barrel, *M* will vary along its length, also the moment of inertia; hence the radius of curvature of the deflection curve will also vary.

Radius of Curvature of the Deflection Curve in the Case of a Gun Barrel

In order to determine the deflection of any point *x,y* on the gun barrel, from its original position, it is necessary to express the value of the radius of curvature, *r* in terms of its displacement *x* along the barrel and its deflection *y* from the original axis.

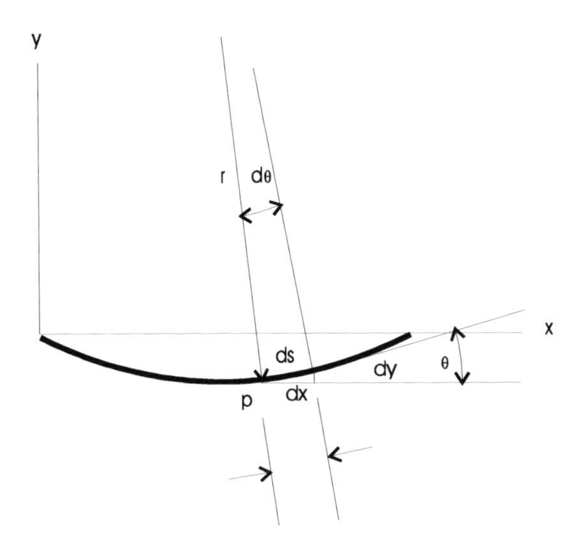

Fig 1.8.1: Radius of curvature in terms of displacement and deflection

From Fig: 1.8.1:

θ being small it can be approximated that:

$tan\ \theta = \theta$ and $ds = dx$.

Hence:

$tan\theta = \theta = \dfrac{dy}{dx}$ for the point P on the deflection curve.

$$\therefore \quad \frac{d\theta}{dx} = \frac{d^2y}{dx^2}$$

Also $rd\theta = ds = dx$

Hence:

$$\frac{1}{r} = \frac{d\theta}{dx} = \frac{d^2y}{dx^2} \quad \text{[1.8.2]}$$

Equating [1.8.1] and [1.8.2] the differential equation of the deflection curve is obtained:

$$EI\frac{d^2y}{dx^2} = M \quad \text{[1.8.3]}$$

By consecutive differentiation of Equation [1.8.3] further results are obtained:

$$\frac{d^3y}{dx^3} = \frac{F}{EI} \quad \text{[1.8.4]}$$

F: shearing force and the shearing force F is the rate of change of the bending moment M with respect to x, given by:

$$\frac{dM}{dx} = F$$

And:

$$\frac{d^4y}{dx^4} = \frac{w}{EI} \quad \text{[1.8.5]}$$

w: weight intensity in weight units /length units at a distance x from the origin, given by:

$$\frac{dF}{dx} = w$$

Relationship between Shear Force and Bending Moment

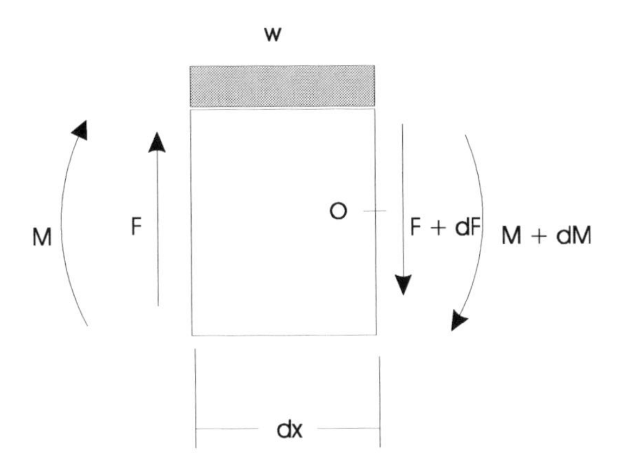

Fig 1.8.2: Relationship between bending moment, shear force and weight intensity

Consider an element of length dx of a beam subject to any kind of transverse load. The shearing force F acts on the left of the element and increases to $F + dF$ on the right hand side of the element. Similarly, the bending moment M on the left side increases to $M + dM$ on the right hand side of the element. Since dx is small the applied load may be taken as uniform and equal to w force units per unit length over the length dx. Taking moments about the point O:

$$M\text{-}(M + dM) + Fdx + wdx\frac{dx}{2} = 0$$

Or:

$$dM = Fdx + \frac{1}{2}(dx)^2$$

Neglecting $(dx)^2$:

$$F = \frac{dM}{dx}$$

Relationship between Weight Intensity and Bending Moment

Considering forces transverse to the beam axis:

$$wdx + F - (F + dF) = 0$$

Or:

$$w = \frac{dF}{dx}$$

Practical Aspects of Barrel Deflection

If the barrel is supported on slide bearings on the cradle, the breech ring itself supports the breech end of the barrel on the cradle. The barrel may be considered a continuous beam, loaded by the weight of the breech assembly, as a point load, at the breech end, with over hang beyond the front support, or a cantilever fixed at the breech end, with a vertical reaction and balancing couple at the breech end and an upward reaction at the front support. The determination of the reactions and the bending moment in such statically indeterminate cases may be solved by more involved methods, such as the use of singularity functions.

A gun barrel always has variations in weight intensity and moments of inertia of various sections along its length. A practical solution is to break up the barrel into sections, which conform to standard shapes when determining the moments of inertia. The reactions at the supports, as also the varying weight distribution resulting in variation in moment of inertia are taken into account.

Any sag between the breech end and the front support is not manifested beyond the front support as the front support is unyielding; and the deflection may be considered from the front support to the muzzle end of the barrel.

General Case of Deflection of a Gun Barrel by Double Integration

For simplicity and in order to illustrate the methodology, the case of a barrel with uniform dimensions and without barrel attachments, like muzzle brakes or fume extractors and with uniform weight distribution is investigated in Example 1.8.1. The barrel is taken as a cantilever, fixed at the front support in a ring bearing, with a length l beyond the front support and of uniform load distribution w/unit length along its protruding length.

Example 1.8.1

Given the following data pertaining to the gun barrel shown in Fig 1.8.3 below:

E: $2*10^{11}$ N/m^2
w: 2412 N/m
l: 3.270 m
Mass density of steel: 7851.45 Kg/m^3
Inner radius of barrel: 0.061m
Outer radius: 0.117 m.

Compute the bending moment and the deflection of the gun barrel.

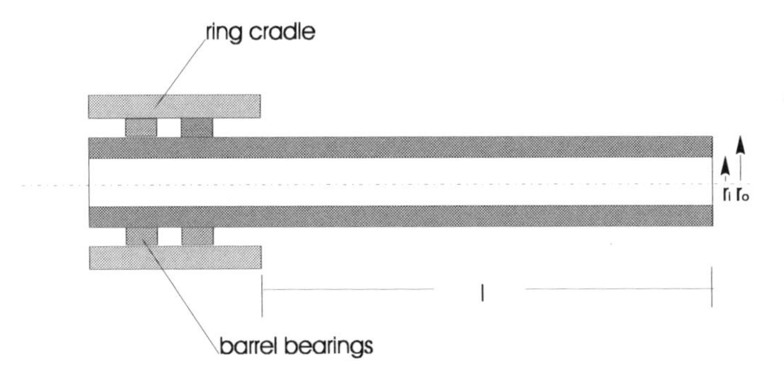

Fig 1.8.3: Gun barrel of Example 1.8.1 as a cantilever

Solution: Example 1.8.1

Deflection of the Gun Barrel

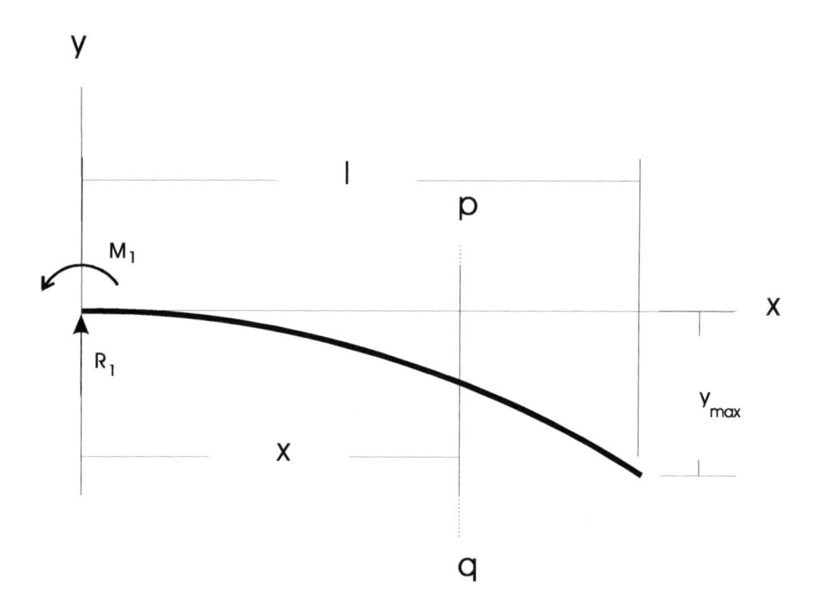

Fig 1.8.4:: Deflection of a gun barrel of Example 1.8.1 by method of double integration

The gun barrel is considered a cantilever, with a uniformly distributed load of w weight units per unit length over its unsupported length and fixed at the breech end.

The vertical reaction at the breech end is:

$$R_1 = wl$$

The balancing moment exerted by the support is:

$$M_1 = \frac{wl^2}{2}$$

The bending moment at any cross-section pq at a distance x from the breech end is:

$$M = w\frac{x^2}{2} - wlx + w\frac{l^2}{2} \quad \text{[1.8.6]}$$

Using Equation [1.8.3]:

$$EI\frac{d^2y}{dx^2} = M = w\frac{x^2}{2} - wlx + w\frac{l^2}{2}$$

A first integration gives:

$$EI\frac{dy}{dx} = w\frac{x^3}{6} - wl\frac{x^2}{2} + \frac{wl^2}{2}x + C_1 \quad \text{[1.8.7]}$$

From the condition that, at the supported end:

$x = 0$ and $\frac{dy}{dx} = 0$

We get:

$C_1 = 0$

A second integration gives:

$$EIy = \frac{w}{24}x^4 - \frac{wl}{6}x^3 + \frac{wl^2}{4}x^2 + C_2$$

Or:

$$y = \frac{wx^2}{24EI}\left(x^2 - 4lx + 6l^2\right) \qquad [1.8.8]$$

From the condition that, at the supported end:

deflection $y = 0$ and $x = 0$

We get:

$C_2 = 0$

From Equation [1.8.8] the deflection along the barrel length can be computed.

Bending Moment

The bending moment is computed using Equation [1.8.6]

Computer Programme 1.8.1

```
% programme to compute bending moment of gun barrel
a=.061% inner radius
b=.117%outer radius
E=2*10^11%youngs modulus
row=7833%density of steel
l=3.270%barrel length from front support
w=2412%weight intensity
x=(0:.1:3.270)
M=w.*x.^2./2-w.*l.*x+w.*l.^2./2 % bending moment
plot(x,M)
title('Bending moment of a gun barrel')
xlabel('barrel length');ylabel('bending moment')
```

Results: Example 1.8.1

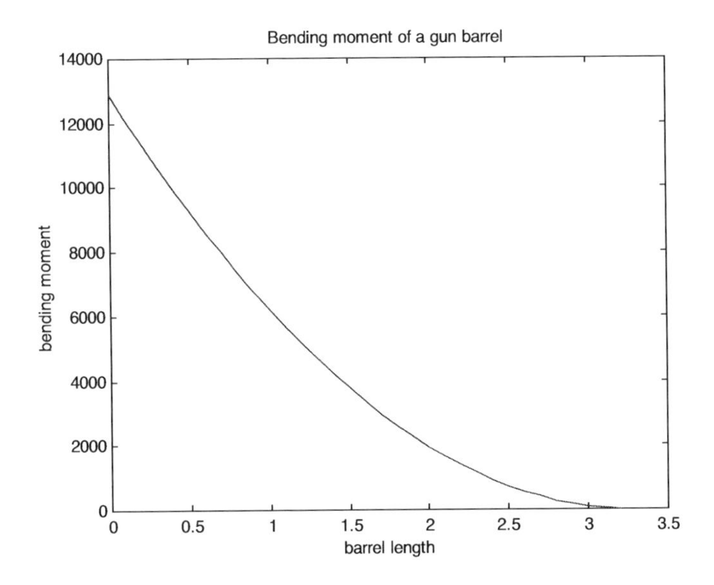

Fig 1.8.5: Bending moment of gun barrel of Example 1.8.1

Deflection of the Gun Barrel of Example 1.8.1

The deflection of the gun barrel is computed as follows:

- The moment of inertia of the section of the barrel about the neutral axis is computed from standard formula.
- The deflection is computed using Equation [1.8.8]

Computer Programme 1.8.2

```
% programme to compute deflection of gun barrel
a=.061% inner radius m
b=.117% outer radius m
E=2*10^11 % youngs modulus
```

```
row=7851.45 % mass density of steel kg/m^3
l=3.270 % barrel length m
w=2412 % weight intensity N/m
x=(0:.1:3.27)
m=pi.*(b.^2-a.^2).*(x).*row  %  mass of length x
I=pi./2.*(b.^4-a.^4) % m of i
y=-(w.*x.^2./(24.*E.*I).*(x.^2-4.*l.*x+6.*l.^2)) % deflection
plot(x, y)
title('Deflection of a gun barrel')
xlabel('barrel length');ylabel('deflection m')
```

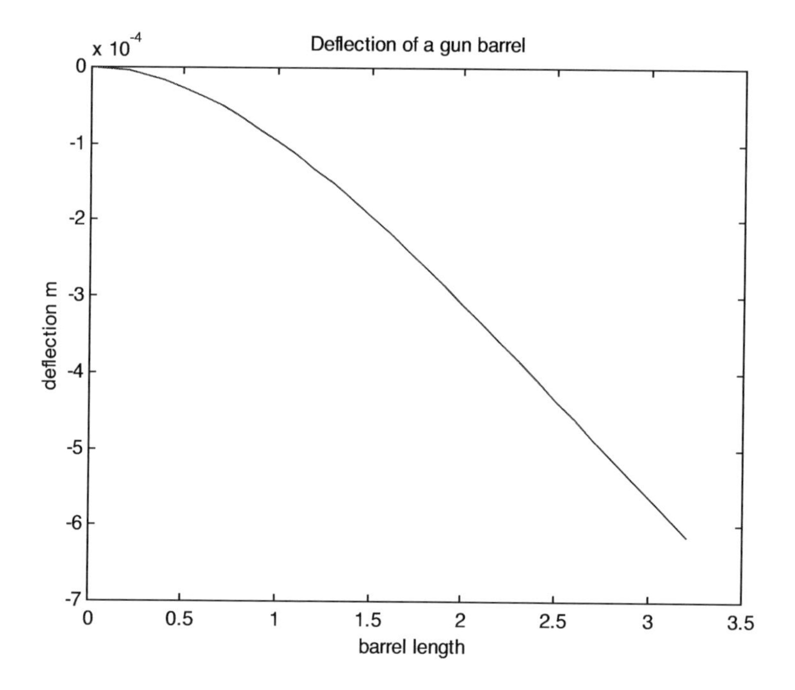

Fig 1.8.6: Deflection of gun barrel of Example 1.8.1 along its length

2

Breech Assemblies

2.1: Breech Rings and Breech Mechanisms

Breech Assemblies

The breech assembly is the component of the gun which shuts the breech end of the barrel, to be more specific, the chamber, during firing and until such time as the pressure inside the chamber is deemed to have fallen to a safe level for the breech to be opened.

Functions of the Breech Ring & Breech Mechanism

The functions of the breech mechanism are as follows:

- To close and lock the breech of the gun.

- To closely support the cartridge case in the case of separate QF ammunition or the full round in the case of fixed ammunition. This implies that the cartridge case must be tightly squeezed between the front detent surface and the breechblock. The cartridge case being relatively thin and of ductile material must not be allowed to expand unrestricted as this will result in its rupture.
- The breech assembly should receive the rearward firing impulses and absorb them without permanent deformation.
- To house the firing mechanism and firing safety arrangements.
- The breech assembly must necessarily incorporate the manual or semi automatic gearing for opening the breech, extraction and ejection of the empty cartridge case, arming of the firing mechanism and closing and locking of the breech.
- In BL guns, to incorporate the means of obturation.
- In guns with breech screws to incorporate the gearing for rotating the screw, axial withdrawal and swinging clear of the screw along with mandatory stops to limit movement and safety locking devices.
- The breech assembly need also contribute to balancing of the gun.

Characteristics of Breech Assemblies

Breech assemblies should ideally possess the following characteristics.

- Reliability and durability.
- Safety of operation.
- Permit quick and easy loading.
- Simplicity of design, manufacture maintenance & repair.
- Accessibility and easy replacement of parts.

Components of the Breech Assembly

The breech end of a gun consists of a breech ring, which is akin to a doorframe into which a sliding door can be slid, or again akin to the neck of a bottle, which can be closed by a threaded bottle stopper. The sliding door in a more elaborate form is called a sliding block mechanism and the stopper in similar circumstances is called a breech screw mechanism.

Breech Rings

The construction of a breech ring depends on the type of breech mechanism it houses. The breech mechanism can be one of two types: the sliding block or the screw type.

Breech Rings for Breech Assemblies with Breech Screws

In breech assemblies with breech screws, the breech ring is circular in shape and is shrunk onto the barrel. The ring and screw engage by means of interrupted threads, which afford easy dismantling.

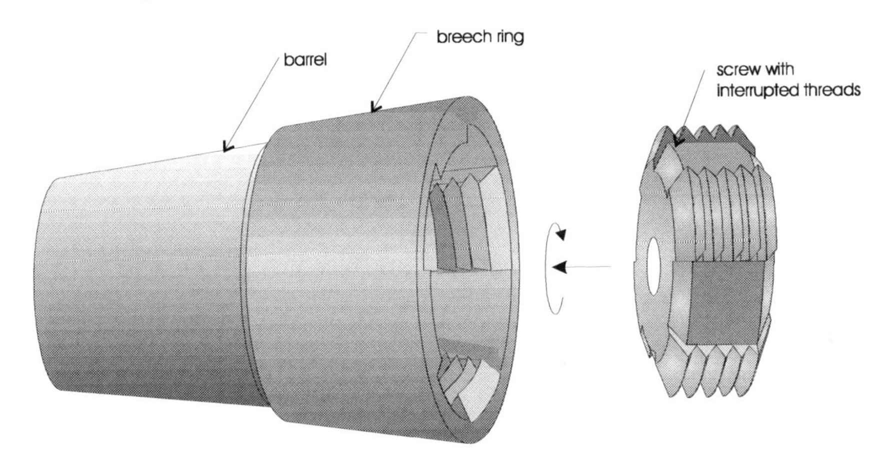

Fig 2.1.1: Breech ring and screw assembly

Breech Rings with Sliding Block Mechanisms

Breech rings with sliding block mechanisms may be of the open or closed jaw type. Both categories are described with illustrations further in the text.

Open Jaw Breech Rings

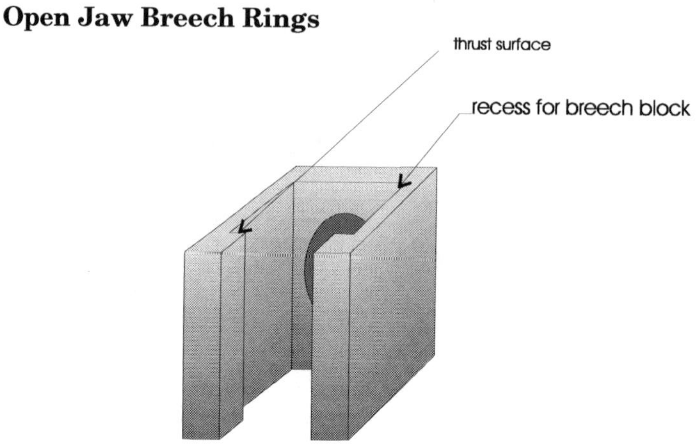

Fig 2.1.2: Open jaw breech ring

Constructional Features of Open Jaw Breech Rings

- The breech ring is hollowed transversely to accept the block.
- The block is provided with ribs working in grooves in the jaws of the ring.

- The ribs guide the block in opening and closing and transmit the firing stresses from the base of the cartridge case to the breech ring.
- Since the back of the ring is open, the firing mechanism and gearing may project from the rear face of the block.

Advantages of Open Jaw Breech Rings

- This type of ring is simpler to manufacture than the closed jaw type.

- It offers access to the firing mechanism without disassembly of the block from the ring.
- It is possible to manually cock the firing mechanism.

Disadvantages of Open Jaw Breech Rings

- The open jaw breech ring has to be inherently strong since the jaws of the ring must be sufficiently rigid to prevent splaying outwards under the influence of bending stresses during firing. The rear thrust surfaces on the block may be beveled and the grooves in the ring dovetailed, to counter act the splaying tendency.
- For a given thrust, this type of ring is thicker, hence heavier than the closed jaw type.
- This design is dimensionally unstable. The dimensions and tolerances existing at the time of manufacture are prone to distortion with use over a period of time due to stress.

Closed Jaw Breech Rings

These are also called tied jaw breech rings. Here the jaws of the breech ring are connected together by a tie.

Constructional Features of Closed Jaw Breech Rings

- In this type of breech ring, the breechblock recess is cut transversely through the ring.
- The rear face of the ring is cut away, partially, to permit loading, yet reducing the overall travel of the breechblock in the open position.
- The rear face of the breechblock bears against the inner rear face of the block recess in the breech ring to which it transmits the firing load.
- One or more pairs of guide ribs may be provided on the blocks, which mate with in grooves in the ring to facilitate the movement of the block within the ring. The ribs also serve to distribute the firing load over a larger surface area, thereby reducing the stresses which the ring is subject to on firing.

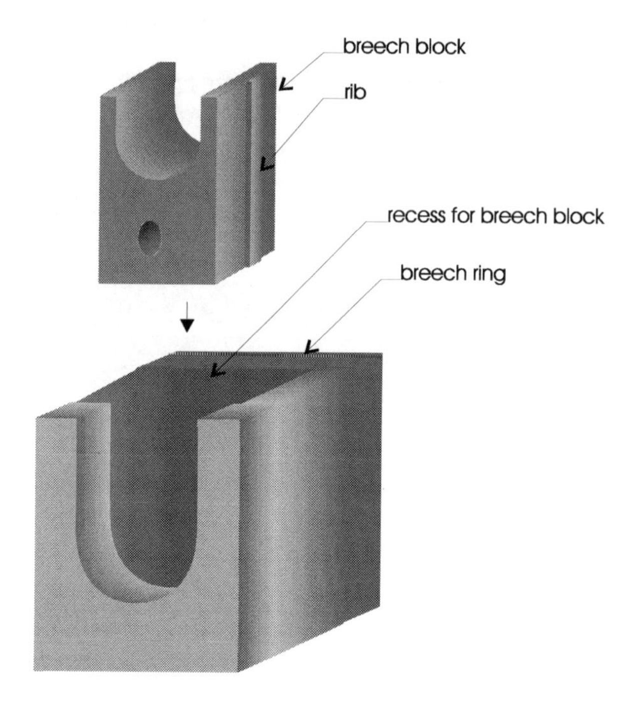

Fig 2.1.3: Closed jaw breech ring

Advantages of Closed Jaw Breech Rings

- This type is inherently stronger than the open jaw type and hence is lighter for a given thrust.
- This type of mechanism has better dimensional stability than the open jaw type of ring.

Disadvantages of Closed Jaw Breech Rings

- This type of breech ring entails more manufacturing processes as also access to the interior of the breech ring is restricted. It is hence more difficult and costly to manufacture as compared to the open jaw type.

- It entails a firing mechanism completely enclosed in the block again complicating manufacture and restricting freedom of design and access to the firing mechanism.

Thrust Surfaces

Open Jaw Breech Rings

If there is one guide rib on each side of the block, the pair forming a single thrust surface the block may be described as single thrust surface block. Further pairs of ribs give double and then multiple thrust surface blocks.

Closed Jaw Breech Rings

With single thrust blocks, the whole thrust is taken on the rear face of the block, the guide serving simply to give rearward travel when block opens. If however guide ribs are also in contact on firing, the rearward thrust is divided between the guide ribs and rear face. The breechblock then becomes a double thrust block. If there are several pairs of guide ribs designed to take part of the thrust it becomes a multi-thrust breechblock.

Bearing Pressure

When the breechblock is in the closed or firing position, it bears against the thrust surfaces of the breech ring. The pressure acting on these surfaces during firing is called bearing pressure. The bearing pressure should be below the yield stress of the material. However, to contain the size of the breech assembly, some local yielding of the more highly stressed parts is acceptable so long as jamming of the mechanism does not occur.

Advantages of Multi Thrust Surfaces

The firing stresses are distributed over the breech ring and not concentrated at a single section.

Disadvantages of Multi Thrust Surfaces

To ensure that each surface takes a proportion of the load, the mating surfaces must be close fitted. The tolerances are extremely small during production by machining. This slows production, demands skilled labour and precludes the interchangeability of blocks and breech rings.

2.2: Stresses in Breech Rings

Stresses on Firing

On firing, the propellant gas force acts on the base of the cartridge case, which in turn exerts a thrust on the face of the breechblock; this thrust is conveyed to the breech ring. It is assumed that the thrust is equally shared by the two jaws and is uniformly distributed about the centre line through the thrust surfaces. As a result the jaws tend to splay outwards. Further, as the centre of the breech ring and the bore axis of the gun do not lie on the same line, in the vertical plane, the point of application of the load is above the centre of the breech ring resulting in eccentric loading.

Stresses in Open Jaw Breech Rings

Stresses during Firing

With reference to Fig 2.2.1, the three kinds of stresses, which occur during firing, are analysed in the following paragraphs.

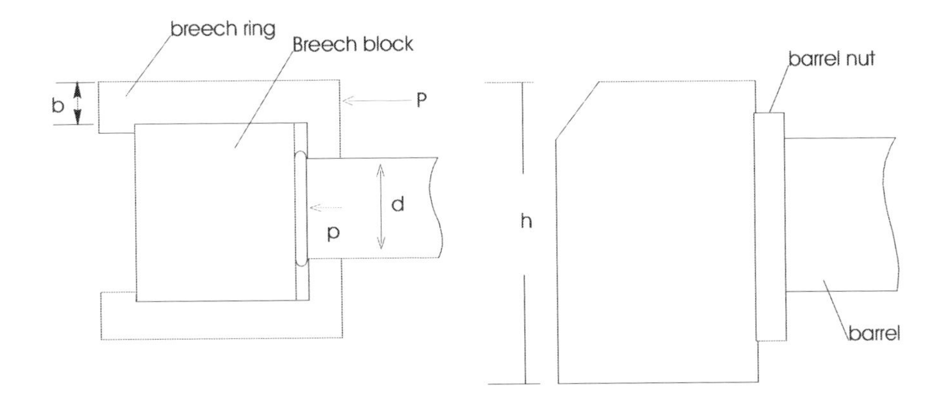

Fig 2.2.1: Forces on an open jaw breech ring

Direct Tensile Stress

Total load transmitted to the breech ring is:

$$2P = p\frac{\pi d^2}{4} \qquad [2.2.1]$$

P: load applied to each jaw
d: maximum internal diameter of the cartridge case
p: maximum propellant gas pressure.

Hence, the direct tensile stress in each jaw is:

$$\sigma_d = \frac{P}{bh} \qquad [2.2.2]$$

b: thickness of the jaw
h: height of the jaw.

Tensile Stress due to Bending

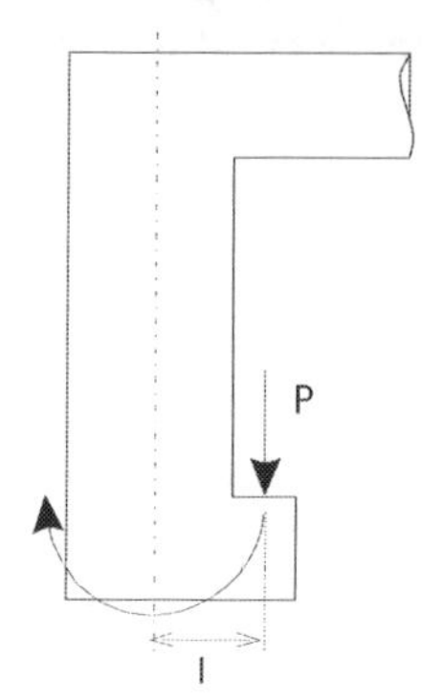

Fig 2.2.2: Tensile stress due to bending

The tensile stress due to bending is determined from the equations for bending stresses of a beam. We have the maximum tensile stress:

$$\sigma_b = \frac{E\frac{b}{2}}{r}$$

E: Young's Modulus
r: radius of curvature of the deflection curve

Also:

$$\frac{1}{r} = \frac{M}{EI}$$

M: bending moment at the cross section under consideration
I: moment of inertia of the section about the neutral axis

Combining the two equations immediately above:

$$\sigma_b = \frac{Pl\frac{b}{2}}{I} \text{ or: } \sigma_b = \frac{Pl\frac{b}{2}}{\frac{b^3h}{12}}$$

$$\sigma_b = \frac{6Pl}{b^2h} \quad [2.2.3]$$

l: distance line of action of force P to neutral axis

Tensile Stress due to Eccentric Loading

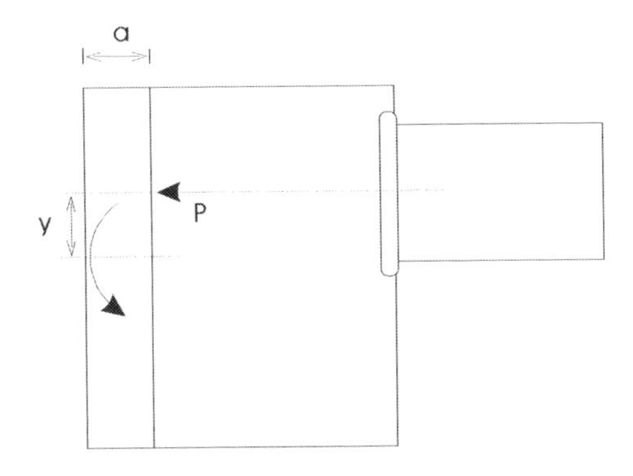

Fig 2.2.3: Tensile stress due to eccentric loading

The bending moment on each jaw due to eccentric loading will be Py

y: offset from axis of bore to line of action of load on thrust surface.

The maximum stress due to this bending moment is:

$$\sigma_e = \frac{Py\frac{h}{2}}{\frac{bh^3}{12}}$$

$$\sigma_e = \frac{6Py}{bh^2} \quad [2.2.4]$$

The net stress is called the nominal stress and is given by the sum of the above stresses mentioned in Equations [2.2.2], [2.2.3] and 2.2.4] above:

$$\sigma_n = \frac{P}{bh}\left(1 + \frac{6l}{b} + \frac{6y}{h}\right) \quad \text{[2.2.5]}$$

Shear Stress

The shear stress calculated across the section of area $a.h$ is:

$$\tau_{AA} = \frac{P}{ah} \quad \text{[2.2.6]}$$

From practical experience this stress should not exceed about 25% of the yield stress of the material.

Correction Factor

From experience it is found that the stresses calculated by the method above are lesser in magnitude than the actual. Hence for open jaw rings the nominal stress computed is multiplied by a correction factor of 1.7.

Stresses in Closed Jaw Breech Rings

The stresses in a closed jaw breech ring are calculated as in the case of the open jaw breech ring, the presence of the tie is ignored. However the correction factor applied is 1.25, which caters for the presence of the tie as also the right angle in place of the re-entrant angle, which is a feature of open jaw rings.

2.3: Sliding Block Breech Mechanisms

Types of Breech Mechanisms

A breech mechanism is a mechanical device for closing the breech end of a gun. Breech mechanisms are of two types: sliding block and breech screw mechanisms

Sliding Block Mechanisms

Breech closing, in this case, is accomplished by sliding a metal block tightly against the base of the cartridge case. The block is then supported and locked in position. The breech ring provides the necessary support. Locking is achieved by an arrangement, which does not allow the movement of the gearing in the direction of opening while the pressure is high. The sliding type of breech mechanism is simpler in operation than the screw type, but makes the gun heavier and so is not preferable for large calibre equipment. Blocks are also restricted to QF guns. On the other hand the block type is readily adaptable to semi and full automatic action, whereas the screw type is not. Here obturation is provided by the metal cartridge case.

Components of a Sliding Block Mechanism

- The breechblock.
- Lever breech mechanism with requisite gearing to open and close the breech manually and if required, automatically.
- A device or an arrangement to hold the block closed i.e. locking.
- A safety arrangement to prevent firing until the breech is fully closed.
- Extractors for removal of the spent cartridge case.
- Various devices to limit movement of the block.

Types of Sliding Blocks

- Horizontal sliding blocks.

    - These require a constant effort to operate both during opening and closing.
    - The effort required at all angles of elevation is constant as the effect of gravitational force is the same regardless of the angle of elevation.
    - The maximum traverse to both sides is reduced to preclude contact of the block, in the open position, with the trail, in positions of maximum traverse. This is achieved by the introduction of traverse stops on the basic structure.

- Vertical sliding blocks.

    - This arrangement requires less effort to open but more effort to close as the block is closed against the force of gravity.
    - The maximum elevation of the gun is restricted or excavations have to be prepared before firing under the line of movement of the breech during recoil. Elevation stops have to be provided on the basic structure, to prevent the block from touching the ground at greater angles of elevation.

- Oblique sliding blocks. This design is obviously a compromise between the advantages and disadvantages of the preceding two types of sliding blocks and finds application in special circumstances, for instance in the restricted space in AFV turrets and fighting compartments.

Breechblocks

Breechblocks are generally rectangular in section and may slide horizontally or vertically in a recess in the breech ring under the action of the LBM and gearing. The sliding block falls under two categories:

- Wedge type breechblock. The rear of the block is slightly wedge shaped and the block wedges tightly against the base of the cartridge case on closing to effectively support it. The thrust during firing is transmitted through the rear of the block to the breech ring and the ribs serve only as guides during the opening and closing movements of the block.

- Rectangular type breechblock. In this design, the block itself is rectangular, but has ribs on its side surfaces, which engage in grooves in the breech ring internal side faces. The firing thrust is transmitted partially by the ribs and partially by the rear of the block to the breech ring. The ribs are slightly inclined towards the front, to give a forward motion to the block (termed as forward travel) on closing in order to firmly contact the base of the cartridge case.

Gearing

Gearing in all cases performs the multiple functions of unlocking and opening of the breech, extraction and ejection, cocking of the firing mechanism and closing and locking of the breech. From the point of view of operation gearing may be of three types:

- Hand operated gearing. The block is operated by a crank, which is connected to the LBM and works in a cam groove or path, cut in the block. The first movement of the LBM does not move the block but simply opens the mechanical safety lock and in some cases withdraws the firing pin. The action of the crank is designed to provide leverage gain to reduce the effort required to manually lift the breechblock. Spring-loaded lugs on the rear face of the extractors are employed to hold the block in the open position. The lugs are displaced by the rim of the cartridge case when the cartridge case is fed into the chamber and the breechblock is then free to close. All sliding blocks have an inherently positive safety feature; the firing pin is not aligned with the primer on the cartridge case until the block is fully closed. With electrical firing circuits, a contact within the breechblock is aligned with a contact on the breech ring only when the block is in the fully closed position.

- Semi automatic gearing. In this case the same general arrangement is used as for hand operation with the additional facility for the semi-automatic action. On run up a cam or bar on the cradle fouls a roller on the breech mechanism, which rotates the shaft on which a crank is mounted. The crank gets rotational movement, which is converted, to vertical movement of the breechblock to open the breech. The rotation of the shaft is also employed to cock the firing mechanism and extract and eject the fired cartridge case as also to compress a closing spring. The extractors being pivoted, this is achieved by lugs on the lower end of the extractors contacting protrusions on surfaces of the breechblock. Mechanical advantage is obtained by varying the effort and load arms of the extractors. The block is now held in the open position by teeth on the rear face of the extractors. As the block is held, so also is the compressed spring prevented from expanding. When a cartridge case is loaded the extractors are forced forward releasing compression of the spring. The energy stored in the spring rotates the shaft and lifts the block, which returns to the closed position under its pressure. A means of hand operation is always incorporated in the gearing for operation of the breech mechanism manually, as for inspection and loading the first round and also in the case of failure of the semi-automatic gear.
- Fully automatic gearing. Here the action on recoil and subsequently run up opens the breech, ejects the empty cartridge loads another round, closes the breech, cocks the firing mechanism and fires the round. Thus the action is fully automatic and the gun will fire continuously as long as rounds are fed into the gun and the sear of the firing mechanism is not allowed to engage the firing pin.

Advantages of Sliding Block Mechanisms

- Sliding blocks as compared to screw mechanisms have the advantages of simplicity of design and manufacture.
- From the operating point of view, they are less complicated and speedier to operate.
- Breechblock mechanisms are safer on closing, as the chance of the loaders hand getting caught between the breechblock and the breech ring while closing it is prevented by the presence of the cartridge case.

Disadvantages of Sliding Block Mechanisms

- Breech mechanisms with sliding blocks require a metal cartridge case to provide obturation. Metal cartridge cases in the ammunition system cause their own complications as concerning manufacture, recovery of fired cases recycling and the logistical problems associated with these aspects.
- Sliding block mechanisms are heavier than screw type of mechanisms for the same thrust. Because the area over which the thrust is distributed is less, the stresses in the material are higher. Naturally the breechblock has to be of sturdier more voluminous and heavy construction.

2.4: Stresses in Breechblocks

Stresses in a Breechblock

A breechblock has large and irregular recesses to house the firing and the opening and closing mechanisms. This leads to an uneven stress distribution in the block. For the purpose of stress analysis the block is assumed to be a simply supported beam carrying a uniformly distributed load over the area covered by the maximum internal area of the cartridge case.

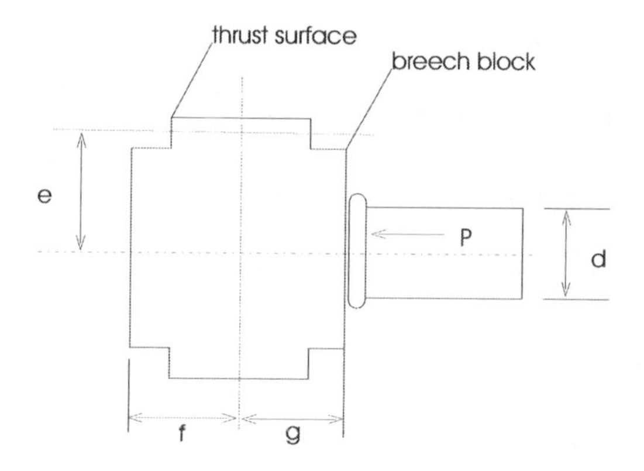

Fig 2.4.1: Bending moment of a breechblock

2P: total thrust applied to the block due to gas pressure
d: internal diameter of the cartridge case
I: moment of inertia of section about the neutral axis
e: distance from axis of bore to centre line of thrust surface
f: distance from rear face of block to neutral axis
g: distance from front face of block to neutral axis

The load distributed over each half segment of the cartridge case is *P*. The maximum bending moment is *P* multiplied by the distance between the centre of pressure of the segment and the centre of the bearing surface:

$$M = (e - \frac{2d}{3\pi}).P = P(e - 0.212d) \quad [2.4.1]$$

The maximum tensile stress is:

$$\sigma_t = M.\frac{f}{I} \quad [2.4.2]$$

The maximum compressive stress is:

$$\sigma_c = M.\frac{g}{I} \text{ ......} \quad [2.4.3]$$

Moment of Inertia of a Breechblock by Mid-ordinate Method

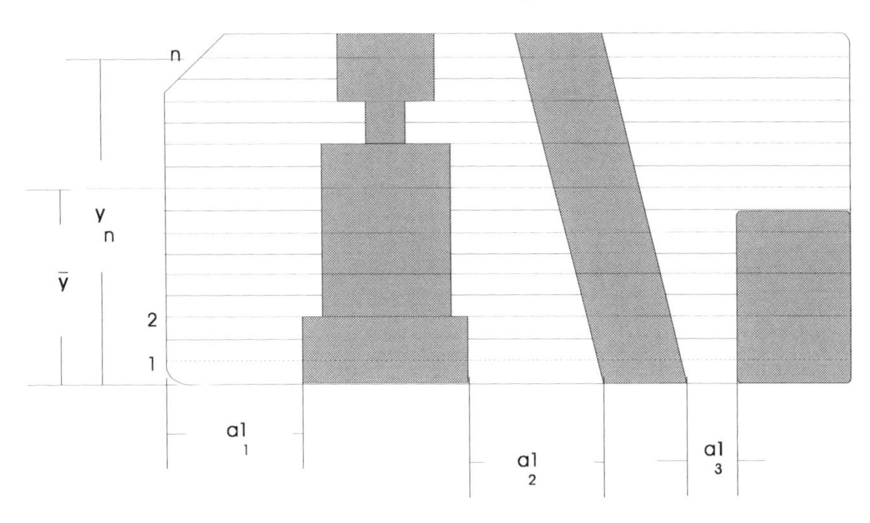

Fig 2.4.2: Centre section of a typical breechblock

Method

The half section of the breechblock is divided into a convenient number of strips of equal width say h. The mid-ordinate of each strip is drawn and numbered. The effective length of each mid-ordinate is measured.

The area of any strip of number n is given by:

$$A_n = (an_1 + an_2 \text{ .... } + an_n)h$$

The next step is to locate the centroidal axis of the area of the section as a whole parallel to the base. If a composite area consists of component areas $A_1, A_2 \ldots\ldots A_n$, at distances $y_1, y_2 \ldots\ldots y_n$ of their centroidal axes, parallel to the base, respectively from the base of the composite area, then the equation of first moments is:

$$(A_1 + A_2 + ...A_n)\bar{y} = A_1y_1 + A_2y_2 + ...A_ny_n$$

$\bar{y}$: distance of the centroidal axis parallel to the base, of the area as a whole, from the base.

Hence:

$$\bar{y} = \frac{\sum_1^n A_ny_n}{\sum_1^n A_n} \quad \text{.......... [2.4.4]}$$

By the parallel axis theorem, the moment of inertia of an area about any axis lying in the same plane is equal to the sum of moments of inertia of the component areas I_n about their parallel centroidal axes plus the sum of the products of each component area and the square of the individual distance of its parallel centroidal axis from the said axis of rotation.

$$I_{zz} = \sum_1^n I_n + \sum_1^n A_ny_n^2 \quad \text{.......... [2.4.5]}$$

I_{zz}: moment of inertia of the area as a whole about the axis parallel and through its base.

Also by the same theorem:

$$I_{zz} = I_{xx} + \left(\sum_1^n A_n\right)\bar{y}^2 \quad \text{.......... [2.4.6]}$$

I_{xx}: moment of inertia of the area as a whole about its centroidal axis, parallel to the axis through the base.

Hence the moment of inertia of the area as a whole about its parallel centroidal axis is:

$$I_{xx} = I_{zz} - \left(\sum_{1}^{n} A_n\right)\bar{y}^2 \quad \text{[2.4.7]}$$

Example 2.4.1

Compute the stresses in a breechblock given the following data:
Maximum gas pressure: 0.227 G Pa
Internal diameter of cartridge case base: 94 mm
Distance axis of bore to centre of thrust surface: 65 mm
Width of the block: 0.122 m

Solution

The centre section of the breechblock is divided into 12 equal strips of 0.01 m each.

- The length of each component of each strip is measured and the effective length found by adding the lengths of all the components within the strip. Length of each strip is listed below:

No of strip	Length l
1	0.146
2	0.147
3	0.147
4	0.117
5	0.111
6	0.107
7	0.095
8	0.090
9	0.085
10	0.087
11	0.079
12	0.068

- The area of each strip is now calculated.
- Thrust P on half segment of the cartridge case is found.
- Maximum bending moment M is calculated using Equation [2.4.1]
- The distance neutral axis to base of breechblock is found with the help of Equation [2.4.4].
- Once the position of the neutral axis is established, the distances to the front and rear faces of the block can be calculated.
- The moment of inertia of the section of the block is found using Equation [2.4.7]
- The maximum tensile stress is computed using Equation [2.4.2]
- The maximum compressive stress is computed using Equation [2.4.3]
- A spreadsheet may be constructed as shown overleaf to conveniently solve the example.

Fig 2.4.3: Half section of breechblock of Example 2.4.1

Spread Sheet: Data & Formulae

Spreadsheet to compute stresses in breech block						
Data						
h	0.01	e	0.065	M	=D4*(D3-0.212*B4)	
d	0.094	P=	=2.27*10^8*B4^2*PI()/8	width	0.122	
n	l	A	yn	area*y	area*y^2	In
1	0.146	=B7*$B$3	=$B$3*A7/2	=C7*D7	=E7*D7	=B7*$B$3^3/12
3	0.147	=B8*$B$3	=$B$3*A8/2	=C8*D8	=E8*D8	=B8*$B$3^3/12
5	0.147	=B9*$B$3	=$B$3*A9/2	=C9*D9	=E9*D9	=B9*$B$3^3/12
7	0.117	=B10*$B$3	=$B$3*A10/2	=C10*D10	=E10*D10	=B10*$B$3^3/12
9	0.111	=B11*$B$3	=$B$3*A11/2	=C11*D11	=E11*D11	=B11*$B$3^3/12
11	0.107	=B12*$B$3	=$B$3*A12/2	=C12*D12	=E12*D12	=B12*$B$3^3/12
13	0.095	=B13*$B$3	=$B$3*A13/2	=C13*D13	=E13*D13	=B13*$B$3^3/12
15	0.09	=B14*$B$3	=$B$3*A14/2	=C14*D14	=E14*D14	=B14*$B$3^3/12
17	0.085	=B15*$B$3	=$B$3*A15/2	=C15*D15	=E15*D15	=B15*$B$3^3/12
19	0.087	=B16*$B$3	=$B$3*A16/2	=C16*D16	=E16*D16	=B16*$B$3^3/12
21	0.079	=B17*$B$3	=$B$3*A17/2	=C17*D17	=E17*D17	=B17*$B$3^3/12
23	0.068	=B18*$B$3	=$B$3*A18/2	=C18*D18	=E18*D18	=B18*$B$3^3/12
sigma	**=SUM(B7:B18)**	**=SUM(C7:C18)**		**=SUM(E7:E18)**	**=SUM(F7:F18)**	**=SUM(G7:G18)**
Dist na to rear face (ybar)				=E19/C19	sigmat=	=$F$3*E21/E23
Izz				=G19+F19	sigmac=	=$F$3*($F$4-E21)/E23
Ixx				=E22-C19*E21^2		

Results

Thrust on the half segment of the cartridge case: $P = 7.88.10^5$ N
Maximum bending moment: 35501.63 Nm
Moment of inertia of centre section about neutral axis: 1.49E-05 m^4
Maximum tensile stress: 1.23E+08 N/m^2
Maximum compressive stress: 1.67E+08 N/m^2

Spreadsheet: Results

Spreadsheet to compute stresses in breech block						
Data						
h	0.010	e	0.065	M	35501.629	
d	0.094	P=	7.88E+05	width	0.122	
n	**l**	**A**	**yn**	**area*y**	**area*y^2**	**In**
1	0.146	0.00146	0.00500	7.30E-06	3.65E-08	1.22E-08
3	0.147	0.00147	0.01500	2.21E-05	3.31E-07	1.23E-08
5	0.147	0.00147	0.02500	3.68E-05	9.19E-07	1.23E-08
7	0.117	0.00117	0.03500	4.10E-05	1.43E-06	9.75E-09
9	0.111	0.00111	0.04500	5.00E-05	2.25E-06	9.25E-09
11	0.107	0.00107	0.05500	5.89E-05	3.24E-06	8.92E-09
13	0.095	0.00095	0.06500	6.18E-05	4.01E-06	7.92E-09
15	0.090	0.00090	0.07500	6.75E-05	5.06E-06	7.50E-09
17	0.085	0.00085	0.08500	7.23E-05	6.14E-06	7.08E-09
19	0.087	0.00087	0.09500	8.27E-05	7.85E-06	7.25E-09
21	0.079	0.00079	0.10500	8.30E-05	8.71E-06	6.58E-09
23	0.068	0.00068	0.11500	7.82E-05	8.99E-06	5.67E-09
sigma	**1.279**	**0.01279**		**6.61E-04**	**4.90E-05**	**1.07E-07**
Dist na to rear face (ybar)				0.05169	sigmat=	1.23E+08
Izz				4.91E-05	sigmac=	1.67E+08
Ixx				1.49E-05		

2.5: Breech Screw Mechanisms

Screw Mechanisms

This type is used for both QF and BL guns and in both cases has the following main parts:

- Breech screw
- Carrier to withdraw and swing the screw clear of the breech
- Means of attachment of breech screw to carrier.
- Means of rotation of the breech screw for engaging with the threads in the breech ring.
- Means of keeping breech screw locked in the open position whilst out of the breech ring.
- Locking device to lock breech screw in the closed position.

In the early days breech screws were used with continuous threads, but they had the disadvantage of slowness in operation. As a result interrupted thread screws were introduced. These are proportionately longer to obtain an equal net thrust surface area as a screw with continuous threads. A number of variations are in use but the one most commonly met in the field is the Welin Screw, other types being cylindrical, front conical rear cylindrical and the tapered screw, in which the largest diameter is towards the muzzle and forms a wedge.

The Welin Screw

This screw has its surface divided circumferentially into segments but these segments have different radii. The circumference of the plain segments varies from one third to one fourth of the total, as against the cylindrical screw where the plain segments occupy half the total circumference. The segment with the shortest radius is plain and the others threaded, so that the plain segments total one third or one quarter of the circumference. The increase in threaded portion allows reduction in length. Also swing clear clearance is greatly reduced. The stress distribution is reasonably good. This reduction of length is important since it allows a shorter breech ring, which is the heaviest part of the gun.

Carriers

These are exactly what the term implies. They carry the bulk of the weight of the breech screw and firing mechanism, particularly when the breech is open. They also allow the withdrawal and swinging clear of the breech screw. These are of two types, the ring and the arm type.

Arm Type Carrier

The carrier is simply an arm hinged to the breech end of the gun, which enables the breech screw to be swung clear of the breech on opening as the screw must not obstruct the loading of the gun. In the majority of cases it has a pintle to which the screw is fastened. It can be used with all forms of breech screws except the cylindrical type.

Ring Type Carrier

These were used with full threaded screws so that the screw had full axial withdrawal before swinging clear.

Attachment of Screws to Carriers

In all cases with screws, the breech is closed by the carrier being swung to the closed position, then the screw is turned, its threads engage with those in the breech ring, until the screw is locked. During this locking action, the screw with regard to carrier rotates and moves forward according to the pitch of the breech threads. The means of attachment of the screw to the carrier must allow for this compound movement.

Gearing

On closing, the breech screw is unlocked from its open position, swung into the breech ring, once inside it is rotated for the threads to engage fully and then locked for safety during firing. All these movements are accomplished by activation of the LBM in one plane only by means of appropriate gearing. The gearing also provides necessary leverage gain for ease of operation.

Devices to Limit Movement

Apart from the gearing some devices to limit movement of the breech screw beyond the extent it is designed for are provided. These do not allow excess rotation of the screw, premature rotation of the screw, swinging out and in of the carrier and movement of the LBM beyond the required limits. They also prevent uncontrolled movement of the screw and carrier.

Obturation in Screw Mechanisms

In BL systems the charge is carried in combustible fabric bag. In such systems it is necessary for the breech mechanism to cater for the sealing of the chamber as also provide a one-way flash vent for the charge to be initiated. The sealing is achieved by a component called an obturator and the flash is conveyed to the charge by what is termed an axial vent.

Obturator

The obturator is a wire woven or asbestos and oil resilient pad shaped like a doughnut, which is contained between metal front and rear protective rings. An inner ring or bearing ring is provided for axial support. The pad sits in a coned seating and is secured to the breech screw by the bolt vent axial and the mushroom head. On firing the mushroom head is forced to the rear by gas pressure thereby squeezing the pad against the front of the breech screw. The pad expands radially and provides a tight seal both at the axial vent spindle as also between the screw and the chamber walls. A typical pad obturator is shown on the next page.

Axial Vent

This is a steel spindle with a head shaped like a mushroom. It has a longitudinal drilling which serves as the flash channel and a chamber similar to that found in small arms. The chamber is closed by a bolt and has a firing mechanism akin to that of a small calibre weapon. The axial vent also incorporates some arrangement to slightly rotate the obturator and free it during the unlocking of the screw.

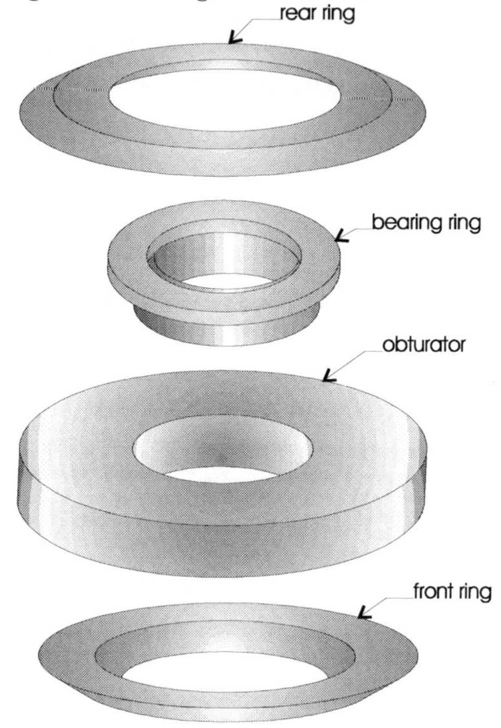

Fig 2.5.1: Pad obturator for screw mechanisms

Advantages of Breech Screws over Breechblocks

- The thrust is spread over the complete area of the threads in the axial direction. Hence the stress is considerably reduced in proportion to the increase in bearing surface.
- The circular shape of the screw and the circumferential distribution of thrust results in a reduction of bending stresses.
- Bending due to eccentric loading is eliminated as the center of pressure is along the axis of the screw.
- Close mating of the threads is possible with accurate machining.
- The overall weight of the breech assembly is reduced.

Disadvantages of Breech Screws over Breechblocks

- Complex to manufacture.
- Slower to operate and not adaptable for semi-automatic and automatic fire.
- Accidents during loading are common.

3

Recoil Systems

3.1: Introduction to Recoil Systems

Development of Recoil Systems

Heavy armament designers have always been plagued with the negative effects of recoil on the stability, aiming, accuracy and rate of fire of the weapon being designed. Elimination or reduction of these effects has been the centre of much attention. Doing away with recoil entirely meant elimination of the recoil system and considerable reduction in the weight of the equipment as a whole.

The first development, based on the principle of conservation of momentum, was that of a gun with a single centrally placed chamber between two opposite facing barrels. The barrel facing forward was meant for the projectile and the one facing backward, a dummy projectile of equal mass. The combustion of the centrally placed propellant propelled the projectile forward and the dummy projectile backward. The masses of the projectiles being equal, the gun remained largely stationary.

The dummy mass was later replaced by a mass of propellant gas ejected rearward from the barrel, at a velocity high enough to equalize the forward and rearward momentums, in spite of the mass difference between the two. This resulted in the barrel remaining stationary. The recoilless gun is based on this principle. Practically, in recoilless guns, the cartridge case is perforated and on firing, allows some of the propellant gas to escape through a nozzle to the rear of the barrel, which increases their velocity. This increase in gas velocity balances the difference in mass between the projectile and the gas, hence the forward and rearward momentum. The recoilless gun succeeds in principle; in practice the blast of hot gases to the rear of the gun is dangerous to soldiers in the vicinity. Also the blast and its scorching effect give away the position of the weapon immediately. Tactically therefore, the employment of the recoilless gun is limited to vehicular mounted light anti armour roles.

Conventional, earlier guns were simply allowed to recoil along with their carriages until they stopped moving, and were then manhandled back into firing position. Only the friction between the wheels and the ground retarded the movement of the gun.

The first attempts at controlling recoil consisted of a platform, the surface which was inclined upwards to the rear. The gun carriage rested on the platform. When the gun was fired, the carriage climbed the slope of the platform against the retarding force of gravity and friction between the carriage and the platform. The friction effect was varied to some extent by coating the surface of the platform with sand or grease.

A further innovation in the development of recoil systems was the use of leaf springs. One end of the leaf spring was attached to the carriage and the other fixed, so that some of the recoil energy was absorbed by elastically stressing the springs. The next improvement was a hydraulic buffer, consisting of a cylinder and piston fixed to the rear of the slide.

The fired gun recoiled until it struck the piston rod and drove the piston into the cylinder against a volume of water, which absorbed the shock. This was subsequently modified by attaching the buffer cylinder to the gun and the piston rod to the cradle. As the gun recoiled, it drove the piston up the cylinder. A hole in the piston head permitted the water to flow from one side of the piston to the other, in a controlled manner. This was the forerunner of the present hydraulic recoil brake. After much modification and improvement, the final addition was the invention of mechanical means of returning the gun to the firing position, by the use of a spring, which stored some of the recoil energy during recoil.

Currently, when a gun recoils, it is braked by the controlled flow of almost incompressible liquid from one side of a piston head to the other within a hydraulic cylinder. At the same time a spring, placed between the gun and the cradle, is compressed and sufficient energy is stored in it. When the energy of recoil has been dissipated and the gun rearward translation ceases, the spring expands and returns the gun to the firing position. Mechanical springs have now been replaced by hydro pneumatic springs of compressed air or nitrogen.

3.2: Recoil and Stability

Free Recoil & Rigid Mounting

It is relevant at the outset to examine two extreme theoretical conditions on firing, firstly that of free recoil i.e. when the gun is allowed unrestricted movement; secondly when the gun is rigidly attached to its supporting structure or cradle.

Free Recoil

When a gun is fired, the gas pressure acts on the face of the closing arrangement of the breech. If the recoiling parts are allowed to recoil freely, they will be accelerated rearward. The entire gas force serves to accelerate the ordnance and the displacement of the recoiling parts continues indefinitely. Further, since there is no arrangement between the static support of the gun and the moving parts to retard the movement of the ordnance, it follows that no reaction force acts at the trunnions.

Rigid Mounting

Assume that the gun is rigidly attached to its fixed support or cradle, which does not move on firing. On firing, there is no displacement; hence, there is no friction force at the slides and no acceleration of the gun. The entire gas force, in this condition, acts at the trunnions and from the trunnions to the supporting structure.

Braking Force and Trunnion Pull

For practical reasons, neither of the two conditions above offers a practical solution. The force acting on the trunnions has to be limited from the point of view of strength and weight of the supporting structure. If the entire gas force is conveyed to the rigid supporting structure through the trunnions, the structure will be impossibly heavy and unwieldy. If the structure is not rigidly fixed to the ground in this condition, it will be displaced and stability of the gun and the supporting structure, as a whole cannot be achieved.

If on the other hand, the gun displaces indefinitely during recoil, retarded only by friction, the length of the slides and the time of recoil would render this solution practically useless. Obviously a finite length of recoil is mandatory.

A compromise is found between the two extreme conditions described above, the gun is allowed to displace, but is brought to a halt in an acceptable distance, by the application of a braking force between the static component, that is the cradle and the dynamic component, the barrel itself. However, unavoidably, the introduction of a braking force to retard the movement of the ordnance results in an equal and opposite force on the trunnions.

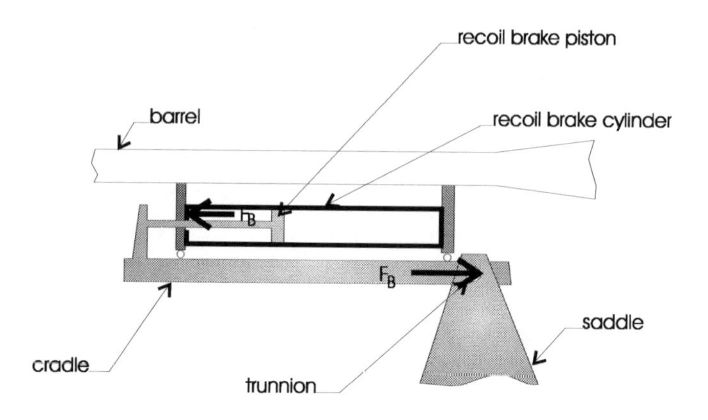

Fig 3.2.1: Braking force and trunnion pull

Recoil and Stability

During retarded recoil, as seen, a force of reaction to the braking force, equal but opposite in direction acts at the trunnions. The moment of this force, about the point of contact of the trails with the ground, called the destabilizing moment, tends to rotate the gun as a whole about this point. This tendency to rotate the gun about the point of contact of the spades and ground is resisted by the weight moment of the gun, as a whole, acting through its center of gravity. When the destabilizing moment exceeds the weight moment in magnitude, the gun becomes unstable. The first manifestation of instability is the lifting of the wheels off the ground. Even minor instability affects the maintenance of aim of the gun and consequently the rate of fire, due to the time required to relay the gun, apart from discomfort to the crew.

Stability Criterion for Recoil Braking Force

It is desirable to bring the ordnance to a halt in a minimum distance, but a definite limit to the magnitude of the braking force, which can be applied, is dictated by the stability condition.

Recoil Length Criterion for Recoil Braking Force

With heavy equipment like tanks, stability of the weapon may not be the deciding factor which dictates the maximum limit of recoil braking force which may be applied. Rather, in such situations it is the recoil length within a confined space which will be the decider. In other words, given the maximum recoil length, the recoil braking force may be computed by dividing the maximum energy of recoil by the given recoil length. The margin of safety here is applied to the maximum recoil length rather than to the maximum braking force.

Stability Criterion for Braking Force

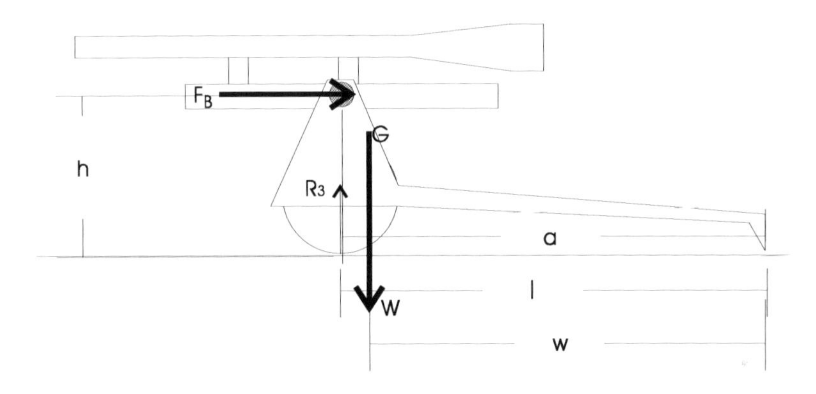

Fig 3.2.2: Destabilizing and stabilizing moment

W: weight of gun as a whole.
G: center of gravity of gun as a whole
F_B: braking force equal to the reaction force on the trunnions
R_3: reaction at the wheels
w,l,h,a: moment arms as indicated

With reference to Fig 3.2.2, taking moments about the point of contact of the trails with the ground, the equilibrium condition is:

$R_3l + F_Bh - Ww = 0$

For stability, the limiting condition is that the wheels do not lift off the ground, that is:

$R_3l = 0$

Hence the condition for stability is simply:

$F_B h = Ww$ .... [3.2.1]

Safety Margin

A practical safety margin s is applied to Equation [3.2.1] to ensure stability under varying conditions. The final equation for stability is now:

$F_B\, h < s\, Ww$ .... [3.2.2]

Effect of Elevation of the Gun

When the gun is elevated or depressed, for that matter, the line of action of the force at the trunnions changes in accordance with the angle of elevation of depression imparted.

In case the gun is elevated, by an angle φ the equation of stability becomes:

$F_BhCos\varphi = Ww + F_BaSin\varphi$ .... [3.2.3]

In case the gun is depressed, by an angle φ, the equation of stability becomes:

$F_BhCos\varphi + F_B\,aSin\varphi = Ww$ .... [3.2.4]

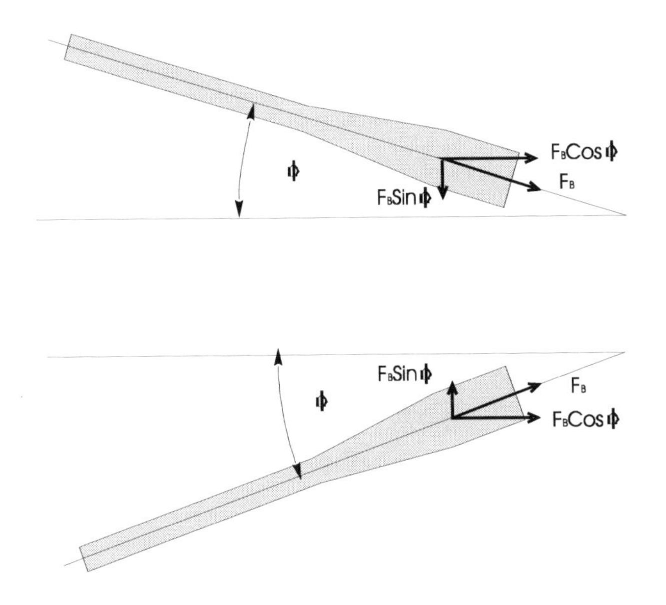

Fig 3.2.3: Effect of angle of elevation on destabilizing moment

With reference to Fig 3.2.3, from the above equations, it is apparent, when the gun is elevated, the vertical component of the reaction to the braking force, acting at the trunnions, adds to the stabilizing moment. When the gun is depressed, the vertical component of the reaction to the braking force acting at the trunnions adds to the destabilizing moment.

3.3: Basic Dynamics of Recoil

Free Recoil

An analysis of free recoil, in which state the gun is assumed to displace under the influence of the gas force without restriction, affords useful information for the more practical application of braked recoil. Considering the simplest case of a gun being fired with the axis of the bore horizontal and with its movement unrestricted.

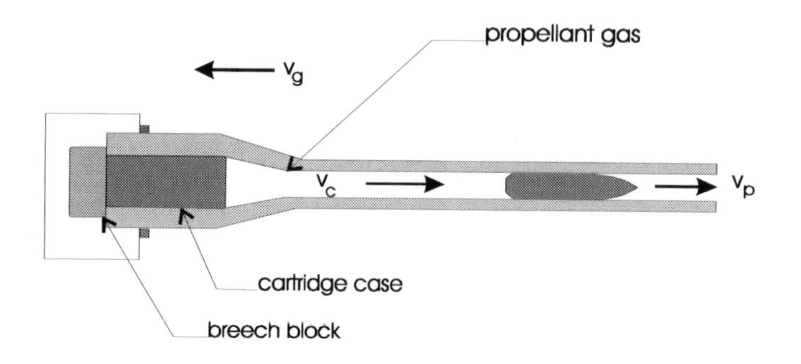

Fig: 3.3.1: Velocities of a horizontal freely recoiling gun

m_g : mass of the recoiling parts

m_p : mass of the projectile

t: time elapsed after the projectile starts moving from rest
m_c: mass of the charge
x_p: distance the projectile has moved in time t
x_g: distance the recoiling parts have moved in time t
A: cross-sectional area of the bore
p : gas pressure at time t
v_c : velocity of the center of gravity of the propellant gas
v_p: velocity of the projectile
v_g : velocity of the recoiling parts

Subscripts:

(0): conditions at shot start
(e): conditions at the instant of projectile exit
(a): the instant at which the gas pressure equals the atmospheric pressure. i.e. when the gas action ceases. The pressure at this instant is approximated as zero.

Since the gun is allowed to recoil freely, the total momentum is constant. Also a reasonable assumption, which holds good up to projectile exit, is that the gas is distributed uniformly throughout the available volume in the barrel.

Relation between Projectile Velocity and Velocity of Recoiling Parts up to Projectile Exit

As the projectile moves with a velocity v_p which is greater than, and opposite in direction to the velocity of the recoiling parts v_g, the center of gravity of the propellant gases will move with a velocity equal to the mean of the difference between the two velocities and in the direction of the greater velocity.

Hence:

$$v_c = \frac{1}{2}\left(v_p - v_g\right)$$

It follows that:

$$m_g v_g = m_p v_p + m_c \frac{1}{2}\left(v_p - v_g\right)$$

Or:

$$\left(m_g + \frac{1}{2} m_c\right) v_g = \left(m_p + \frac{1}{2} m_c\right) v_p$$

Or:

$$v_g = \frac{\left(m_p + \frac{1}{2} m_c\right) v_p}{m_g + \frac{1}{2} m_c} \qquad [3.3.1]$$

In accordance with the assumption that the gas is distributed uniformly throughout the available volume in the barrel, this equation holds good up to the instant of projectile exit.

Hence at the instant of projectile exit:

$$v_{g(e)} = \frac{\left(m_p + \frac{1}{2}m_c\right)v_{p(e)}}{m_g + \frac{1}{2}m_c} \qquad [3.3.2]$$

The relationship between $v_{g(a)}$ and $v_{p(a)}$ has been experimentally found by the Krupps Company to be:

$$v_{g(a)} = \frac{\left(m_p + Km_c\right)v_{p(a)}}{m_g + \frac{1}{2}m_c} \qquad [3.3.3]$$

K: a constant called Krupps constant and varies from 1.5 to 2.5 for different gun tubes.

Because the mass of the charge m_c is small compared to the mass of the recoiling parts m_g, Equations [3.3.2] and [3.3.3] may be approximated as:

$$v_{g(e)} = \frac{\left(m_p + \frac{1}{2}m_c\right)v_{p(e)}}{m_g} \qquad [3.3.4]$$

And:

$$v_{g(a)} = \frac{\left(m_p + Km_c\right)v_{p(a)}}{m_g} \qquad [3.3.5]$$

Maximum Energy of Free Recoil

It is evident that the maximum velocity of the recoiling parts occurs at the instant when the projectile velocity is maximum, which is when the gas pressure is the minimum. The gas pressure is at its least value when it falls to atmospheric, i.e. at condition *(a)*. Therefore, substituting for the velocity of the recoiling parts at the end of gas action the maximum energy of free recoil is:

$$E_{g(a)} = \frac{1}{2} m_g v_{g(a)}^2$$

$$E_{g(a)} = \frac{m_g}{2} \frac{(m_p + Km_c)^2 v_{p(a)}^2}{m_g^2}$$

$$E_{g(a)} = \frac{(m_p + Km_c)^2 v_{p(a)}^2}{2m_g} \quad \text{[3.3.6]}$$

Displacement of the Recoiling Parts up to Projectile Exit

The displacement of the recoiling parts from shot start is computed by integrating Equation [3.3.1] from $t = t_{(0)}$ to $t = t_{(e)}$.

$$\int_{t(0)}^{t(e)} v_g dt = \frac{\left(m_p + \frac{1}{2} m_c\right)}{m_g + \frac{1}{2} m_c} \int_{t(0)}^{t(e)} v_p dt$$

$$x_{g(e)} - x_{g(0)} = \frac{\left(m_p + \frac{1}{2} m_c\right)}{m_g + \frac{1}{2} m_c} \left(x_{p(e)} - x_{p(0)}\right)$$

At the commencement of recoil:

$$x_{g(0)} = x_{p(0)} = 0$$

$$x_{g(e)} = \frac{\left(m_p + \frac{1}{2} m_c\right)}{m_g + \frac{1}{2} m_c} x_{p(e)}$$

The distance traveled by the projectile relative to the bore is $x_{g(e)} + x_{p(e)} = l$; where l the length of the barrel:

Therefore:

$$x_{g(e)}\left(m_g + \frac{1}{2} m_c\right) = \left(m_p + \frac{1}{2} m_c\right)\left(l - x_{g(e)}\right)$$

i.e.

$$x_{g(e)} = \frac{\left(m_p + \frac{1}{2} m_c\right) l}{\left(m_g + m_p + m_c\right)} \qquad [3.3.7]$$

Displacement of the Recoiling Parts and Time Interval after Projectile Exit to End of Gas Action

The general equation of the recoiling parts is given by:

$$m_g \frac{dv_g}{dt} = pA$$

When t' denotes time elapsed after projectile exit, the total time elapsed after shot start is:

$$t = t_{(e)} + t'$$

Accepting D'Valliers hypothesis, that pressure p falls off linearly with time after projectile exit, the pressure falls according to:

$$p = p_{(e)}\left(1 - \frac{t'}{t_{(a)}}\right)$$

At the instant of shot ejection $t' = 0$ and at the end of gas action $t' = t_{(a)}$.

$$m_g \frac{dv_g}{dt'} = Ap_{(e)}\left(1 - \frac{t'}{t_{(a)}}\right)$$

Integrating from $t' = 0$ *to* any time t':

$$m_g \int_{v_{g(e)}}^{v_g} \frac{dv_g}{dt'} dt' = \int_{t_{(e)}}^{t'} Ap_{(e)}\left(1 - \frac{t'}{t_{(a)}}\right) dt'$$

$$m_g\left(v_g - v_{g(e)}\right) = Ap_{(e)}\left(t' - \frac{t'^2}{2t_{(a)}}\right) \quad \text{[3.3.8]}$$

Therefore, when $t' = t_{(a)}$:

$$m_g\left(v_{g(a)} - v_{g(e)}\right) = Ap_{(e)}\frac{t_{(a)}}{2}$$

Using Equations [3.3.4] and [3.3.5] and assuming $v_{p(e)} = v_{p(a)}$

$$m_g\left[\left(\frac{m_p + Km_c}{m_g}\right)v_{p(a)} - \left(\frac{m_p + \frac{1}{2}m_c}{m_g}\right)v_{p(a)}\right] = \frac{Ap_{(e)}t_{(a)}}{2}$$

$$\left(K - \frac{1}{2}\right)m_c v_{p(a)} = \frac{Ap_{(e)}t_{(a)}}{2}$$

$$\therefore \quad t_{(a)} = \frac{(2K-1)v_{p(a)}m_c}{Ap_{(e)}} \quad \text{.................. [3.3.9]}$$

Integrating Equation [3.3.8] above from $t' = 0$ to $t' = t_{(a)}$:

$$m_g \int_0^{t_{(a)}} \left(v_g - v_{g(e)}\right)dt' = Ap_{(e)} \int_0^{t_{(a)}} \left(t' - \frac{t'^2}{2t_{(a)}}\right)dt'$$

Or:

$$m_g\left[\left(x_{g(a)} - x_{g(e)}\right) - \left(v_{g(e)}t_{(a)}\right)\right] = Ap_{(e)}\left(\frac{t_{(a)}^2}{2} - \frac{t_{(a)}^3}{6t_{(a)}}\right)$$

$$m_g\left[x_{g(a)} - x_{g(e)} - v_{g(e)}t_{(a)}\right] = Ap_{(e)}\frac{t_{(a)}^2}{3}$$

$$x_{g(a)} - x_{g(e)} = Ap_{(e)}\frac{t_{(a)}^2}{3m_g} + v_{g(e)}t_{(a)}$$

Using Equations [3.3.4] and [3.3.9]:

$$x_{g(a)} - x_{g(e)} = \frac{(2K-1)v_{p(a)}m_c t_{(a)}}{3m_g} + \left(\frac{m_p + \frac{1}{2}m_c}{m_g}\right)v_{p(a)}t_{(a)}$$

$$x_{g(a)} - x_{g(e)} = \frac{v_{p(a)}t_{(a)}}{m_g}\left[m_p + m_c\left(\frac{2k-1}{3} + \frac{1}{2}\right)\right]$$

$$x_{g(a)} - x_{g(e)} = \frac{v_{p(a)}t_{(a)}}{m_g}\left[m_p + \left(\frac{2k}{3} + \frac{1}{6}\right)m_c\right] \quad \text{.................. [3.3.10]}$$

Hence knowing $x_{g(e)}$ from Equation [3.3.7]

$$x_{g(a)} = \frac{v_{p(a)}t_{(a)}}{m_g}\left[m_p + \left(\frac{2K}{3} + \frac{1}{6}\right)m_c\right] + x_{g(e)} \qquad [3.3.11]$$

The equation immediately above gives the total displacement of the recoiling parts, from shot start to the end of gas action.

3.4: Recoil Forces

Forces before Firing

The forces acting before the gun fires are depicted in Fig 3.4.1 below. The elevation of the gun considered when computing these forces is the maximum elevation that the gun is intended to fire at. If the initial force, called the recuperator force, is sufficient to hold the gun in the run up position at maximum elevation, it will suffice for all other angles of elevation.

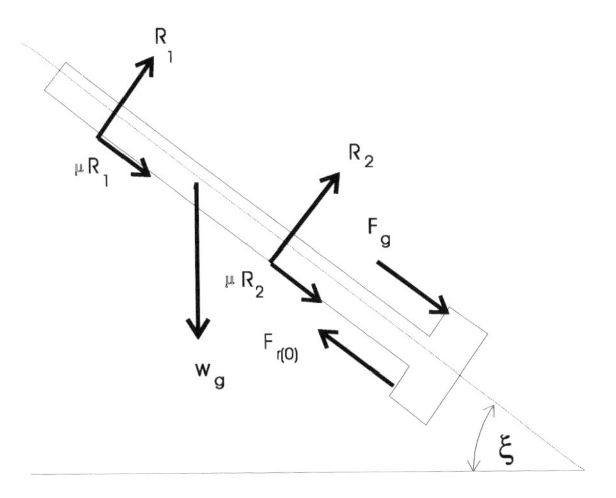

Fig 3.4.1: Forces on a gun before firing

F_r: recuperator force
$F_{r(0)}$: initial recuperator force
R_1, R_2: reactions at the brackets
W_g: weight of the recoiling parts
F_B: braking force
$F_{B(0)} = F_g$: Initial braking force which is equal to the friction force F_g between glands and pistons
μ : coefficient of friction between slides & guides
ξ : maximum elevation angle

Note: For a more detailed treatment on friction force of the glands see Section 9: Seals and Sealing, of this chapter.

Resolving the forces transverse to the gun axis:

$$W_g Cos\xi = R_1 + R_2 \qquad [3.4.1]$$

Resolving the forces along the gun axis, the condition that the gun will remain in the run up position is:

$$F_{r(0)} > \mu(R_1 + R_2) + W_g Sin\xi + F_g$$

Or:

$$F_{r(0)} > \mu\, W_g Cos\xi + W_g Sin\xi + F_g \qquad [3.4.2]$$

If a practical safety factor of λ is allowed, the recuperator force required to hold the gun in the firing position at all angles of elevation is:

$$F_{r(0)} = \lambda(\mu\, W_g Cos\xi + W_g Sin\xi + F_g) \qquad [3.4.3]$$

Dynamics of the Recoiling Parts

When the gun is fired at an elevation φ, the forces acting on the gun are as depicted overleaf:

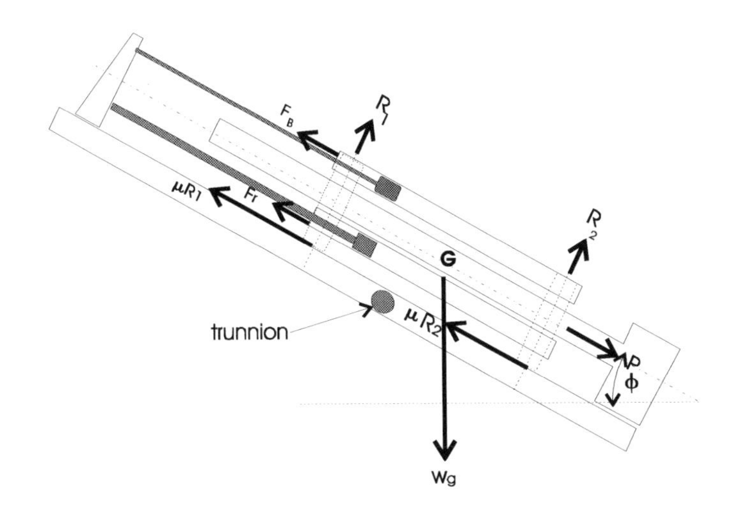

Fig 3.4.2: Forces on recoiling parts of a gun during firing

G: center of gravity of the recoiling parts
P: gas force
ϕ : angle of elevation

Resolving the forces along the axis of the bore:

$$P - F_B - F_r - \mu(R_1 + R_2) + W_g Sin\,\phi = m_g \frac{dv_g}{dt} \quad [3.4.4]$$

The summation of forces on the left hand side of the expression immediately above gives the force accelerating the recoil parts:

$$F_a = m_g \frac{dv_g}{dt} \quad [3.4.5]$$

F_a: acceleration force.

Defining the Maximum Braking Force Permissible for Stability

As seen, a gun, which is stable at zero elevation, will be so at higher elevations, so it is sufficient to consider the case of a gun at an angle of elevation of zero when attempting to fix the maximum value of braking force.

The stability condition with safety margin applied as given by Equation [3.2.2] is:

$$F_B h = sWw$$

Hence the maximum permissible value of braking force consistent with stability is:

$$F_{B\max} = \frac{sWw}{h} \qquad [3.4.6]$$

Where F_{Bmax} includes F_r and also the friction forces.

And:

$F_{r(0)}$ is given by Equation [3.4.3] as:

$$F_{r(0)} = \lambda(\mu W_g Cos\xi + W_g Sin\xi + F_g)$$

Based on the terms reflected in Equation [3.4.6] the following important conclusions can be drawn.

- A reduction in trunnion height h implies better stability for a fixed maximum braking force, or a possible increase in the magnitude of the braking force with a shorter recoil length.
- Stability or magnitude of the magnitude of the braking force may be increased with an increase in the weight of the gun as a whole.
- With an increase in the trail length, stability increases, or, the braking force may be increased without detrimental effect on stability.

As seen earlier, the force on the trunnions is equal and opposite to the braking force, which is comprised of three components, the recuperator force, the gland friction force and the braking force itself. Initially the braking force is minimal, being equal to the friction force of the glands alone. Also the recuperator force is at its minimum value $F_{r(0)}$. In order to have low peak forces it is ideal to maintain the net braking force constant through out the recoil. However, in order to allow the recoiling parts to accelerate, the net braking force must be low at the start of recoil, then rise to a maximum and constant value, which is maintained till the end of recoil. Definition of the maximum braking force permissible from the instant of shot start till the end of recoil is best illustrated with the help of an example.

Example 3.4.1

The following data pertains to a 76 mm gun.

Overall mass of gun: 1200 kg
Trunnion height: .875 m
Safety margin: 150%
Mass of projectile: 6.2 kg
Mass of propellant: 1.08 kg
Length of rifled bore: 2.587 m
Mass of recoiling parts: 402 kg
Muzzle velocity: 680 m/s
Krupps constant: 2
Coefficient of friction: 0.2

From the internal ballistics solution:

Pressure at projectile exit: 43.44 M Pa
Muzzle velocity: 680 m/s

Compute the maximum permissible braking force and plot this force against displacement of the recoiling parts.

Solution

- The maximum permissible braking force is given by Equation [3.4.6].

- In order to allow the recoiling parts to accelerate and gain kinetic energy, the maximum permissible braking force is applied at the end of gas action.

- In order to compute the displacement of the recoiling parts at the end of gas action, it is necessary to calculate the time to the end of gas action with the help of Equation [3.3.9].

- The displacement of the recoiling parts is now computed with the help of Equation [3.3.11]

- The inescapable braking force is given by Equation [3.4.3]

- The braking force is made to rise linearly from its minimum value to its maximum value over the displacement up to the end of gas action. This linear increase is plotted against the displacement from shot start to end of gas action.

- In order to find the recoil length from end of gas action to the end of recoil, the following procedure is adopted:

    - The maximum energy of free recoil is computed using Equation [3.3.6].
    - In order to bring the recoiling parts to a halt, the recoil brake, recuperator and the friction forces have to perform an equivalent amount of work on the recoiling parts.
    - The work performed on the recoiling parts up to the end of gas action is the sum of the friction force times the distance moved by the recoiling parts up to the end of gas action, plus the work done by linearly increasing the braking force from the initial recuperator force to the maximum permissible braking force. The maximum permissible braking force being

the maximum braking force from the stability point of view multiplied by the safety margin.

- The recoil length from end of gas action to the end of recoil is now the remainder energy of the recoiling parts, divided by the maximum permissible braking force.

Computer Programme

```
% programme to compute braking force during the recoil cycle
% data
W=1200.*9.81 % overall weight of gun N
w=3 % distance line of action to point of contact of trails m
H=.875 % trunnion height m
s=2./3 % safety margin
mp=6.2 % mass of projectile kg
mc=1.08 % mass of propellant kg
l=2.587 % length of rifled barrel or shot travel m
mg=402 % mass of recoiling parts kg
wg=402.*9.81 % weight of recoiling parts N
vpa=680 % velocity of projectile is approximated as muzzle velocity of
projectile m/s
k=2 % Krupps constant
D=.076 % calibre m
A=pi./4.*D.^2 % area of bore m²
pe=43444241.13 % gas pressure at projectile exit from internal ballistics
solution Pa
mu=.2

% maximum braking force for stability
F_stab=W.*w./H

%maximum permissible braking force
FBmax=W.*w.*s./H % maximum permissible braking force

% travel of the gun at shot exit
xge=(mp+.5.*mc).*l./(mg+mp+mc)
```

```
% travel of gun at end of gas action
ta=(2.*k-1).*vpa.*mc./(A.*pe)
xga=(vpa.*ta./mg).*(mp+(2.*k./3+1./6).*mc)+xge

% to compute displacement of recoiling parts till end of recoil
Ea=(mp+k.*mc).^2.*vpa.^2./(2.*mg) % recoil energy

% friction force or initial braking force
FB0=mu.*wg

%Work done by friction force and braking force up to end of gas action
Efb=(FBmax-FB0)./2.*xga+FB0.*xga

%remainder energy of recoiling parts at instant of end of gas action
Ega=Ea-Efb

% displacement of recoiling parts from end of gas action to end of recoil
xf=Ega./FBmax+xga

%Plot of maximum permissible braking force vs displacement
x=linspace(0,xf,100)
m=(FBmax-FB0)./xga % slope FB0 to FBmax
x0_a=linspace(0,xga,22)
y0_a=m.*x0_a+FB0
xa_f=linspace(xga,xf,100)
plot(xa_f,FBmax,x0_a,y0_a,x,FBmax,x,F_stab)
```

Results

Maximum permissible braking force: 2.6907e+004 N

Time to the end of gas action: 0.0112 s

Recoil length: 0.1905 m

Minimum braking force: 788.7240 N

Recoil length from end of gas action to the end of recoil:

- Maximum energy of free recoil: 4.0195e+004 J
- Recoil length; end of gas action to end of recoil: 1.5863 m

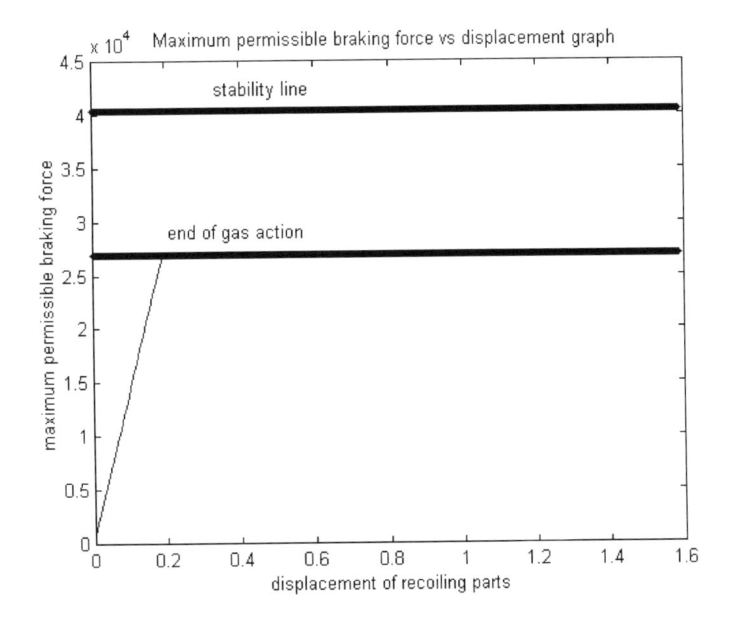

Fig 3.4.3: Maximum permissible braking force-displacement graph

3.5: Functions of a Recoil System

Dissipation of Recoil Energy

As mentioned earlier, if a gun is attached rigidly to its supporting structure, the supporting structure will be subject to a heavy impulse on firing. To withstand such impulse loading, the supporting structure must be strong enough. In the case of small arms, the human body provides the support, which absorbs the firing impulse. In the case of heavy weapons, the strength of the material and size of the base, required to prevent failure of the material of the structure or overturning of the gun, if rigidly attached to the supporting structure, is enormous. Hence a recoil mechanism is incorporated between the gun and the supporting structure.

The recoil system initially allows the recoiling parts to accelerate under the force of the propellant gases. The energy of the propellant gases gets converted to kinetic energy of the recoiling parts. The motion of the recoiling parts is then retarded by a braking force acting over a finite length. This retarding force consists of two components; one, the friction of the slides and the glands, two, the resistance of the recoil mechanism.

The product of recoil force and recoil distance being recoil energy; is equated to the retarding force times the distance of recoil. In other words, the energy of recoil has to be equaled by the work done by the recoil brake to bring the gun to a halt in a fixed distance, less of course, the energy dissipated by friction and energy stored for returning the gun to the firing position.

The magnitude of the retarding force can be controlled by increasing, or decreasing, the distance over which the retarding force acts. In other words by increasing, or decreasing the recoil length. The net energy of recoil is dissipated or utilized in the following ways:

- Deflection of the structure.
- Gun slide and gland friction.

- Dissipation by the recoil mechanism as heat.
- Storage in the recuperator to return the gun to the firing position.

Operating Principle

All recoil mechanisms operate on the basic principle of:

- Providing a controlled resistance within a fixed distance to retard the motion of the recoiling parts. The area under the force-distance curve represents the recoil energy. A rectangular curve will yield the least peak force; hence the net resistance to recoil should be nearly constant. However a rectangular curve is not practical at the beginning, as the recoiling parts should be initially free to accelerate. Also there always is some minimum recuperator force present to hold the ordnance in the firing position at all angles of elevation.
- Returning the recoiling parts to the firing position.
- Holding the recoiling parts in the firing position at the maximum angle of elevation.
- While executing the first two functions given above, ensuring the stability of the equipment both during recoil and counter recoil where stability is a criterion.
- While executing the first two functions given above, ensuring the length of recoil is not exceeded in equipment where stability is not the criterion.

Recoil Cycle

As soon as the gun is fired, the propellant gas pressure accelerates the recoiling parts backwards. This motion is resisted by the inertia of the recoiling parts, friction and the recoil mechanism. At the instant of firing only recuperator and friction forces are present. After motion begins, these are augmented by the hydraulic throttling or braking force. Due to compression of the spring or gas the recuperator force increases gradually, storing the energy needed for counter recoil. At the completion of the recoil stroke, the recuperator begins to return the recoiling parts back to the firing position. High counter recoil velocity is not desirable in heavy weapons; hence the velocity of counter recoil is controlled. Just before completion of the counter recoil stroke, the moving parts contact the buffer, which cushions the final movement of the recoiling parts and stops them as they reach the firing position.

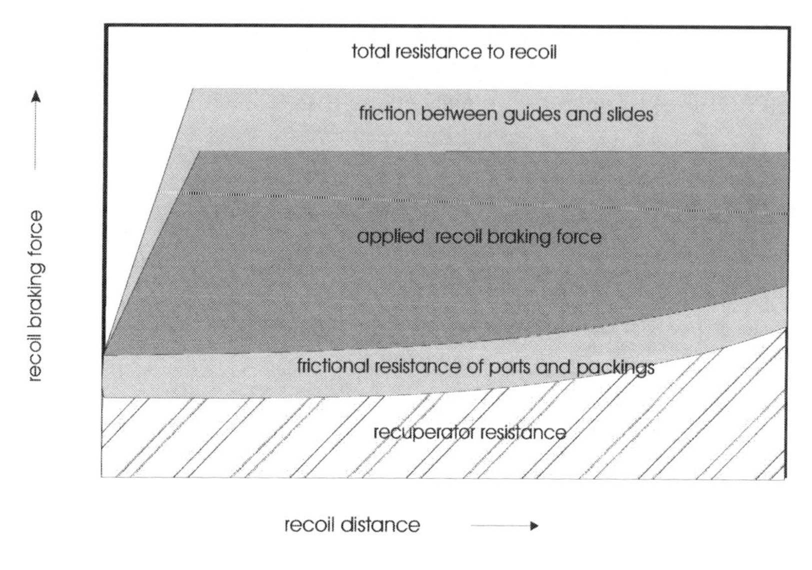

Fig 3.5.1: Operating characteristic of a recoil system

Magnitude of Braking Force

As evident from Fig 3.5.1, the recoil braking force is the major component of the retarding force throughout the recoil length. However, its magnitude varies in accordance with the recuperator force in order that the net braking force is kept constant after the end of gas action. This is an important aspect of recoil system design.

3.6: Counter Recoil

Counter Recoil

Counter-recoil is the motion of the recoiling parts from the fully recoiled position back to the firing position. The requirement in the final stages of counter recoil is to bring the gun to a smooth halt without affecting its stability. It can be appreciated that although the reaction to the force exerted by the recuperator at the trunnions is much smaller than the force on the trunnions during recoil, the weight moment on which the stability depends during counter recoil is also much less than the weight moment during recoil. This is due to the fact that the axis about which the gun will tip over is now the line passing through the points of contact of the wheels with the ground. This axis is now much closer to the line of action of the centre of gravity of the gun, hence a reduced moment arm of the stabilizing moment.

The period of counter recoil is influenced by the recuperator force, based on the energy stored in the recuperator during recoil, the retarding force of the recoil brake in which the motion of some or all the fluid is retarded, friction forces attributed to the slides and packings and the weight component of the recoiling mass along the reversed direction of motion as mentioned.

Notation used in Fig 3.6.1:

F_{Bcr} : net braking force during counter recoil
F_r : recuperator force
ϕ : angle of elevation
v_{cr} : velocity of the counter recoiling parts
R_1 and R_2: reactions at the supports
R_4: reaction at the point of contact of trails
μ : coefficient of friction between the slides and guides

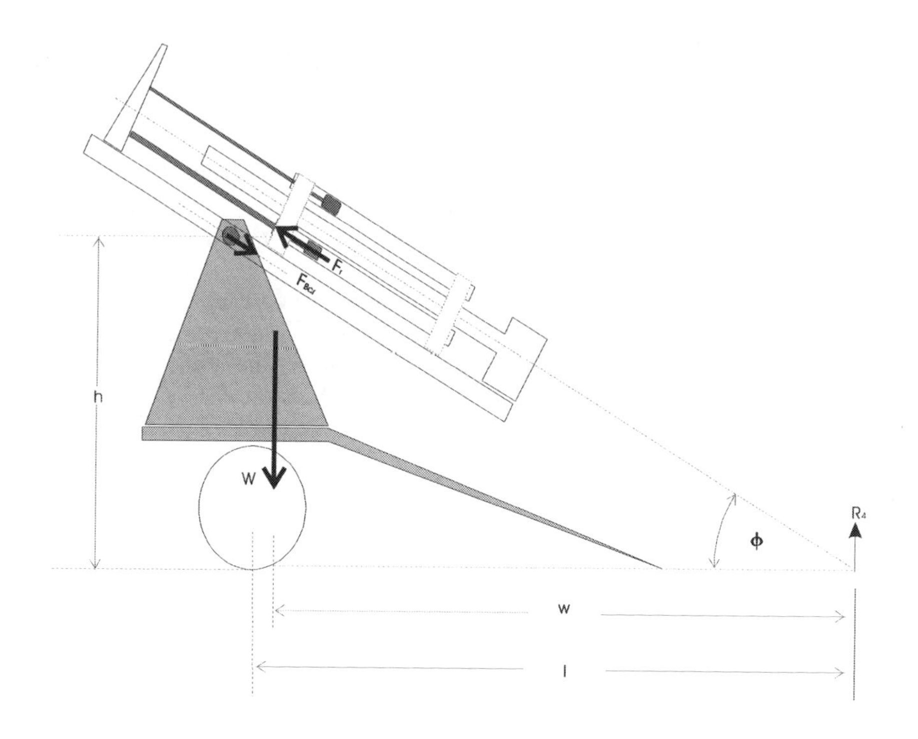

Fig 3.6.1: Forces acting on the gun during counter recoil

Dynamics of the Counter Recoiling Parts

Resolving the forces acting on the recoiling parts parallel to the gun axis:

$$F_r - F_{Bcr} - W_g Sin\phi = m_g \frac{dv_{cr}}{dt} \qquad [3.6.1]$$

Stability during Counter Recoil

The value of the net braking force during counter recoil is again limited by stability conditions. If the recuperator force is large at the end of recoil when the recoiling parts contact the cradle cap, the gun will nose over. Also when the recoiling parts are braked during counter recoil, a reaction force equal and opposite to the braking force acts at the trunnions giving rise to a destabilizing moment.

During counter recoil, if the gun is unstable it will rotate about the axis through the point of contact of the wheels with the ground and tip over, muzzle first. It is important, while restricting the velocities of counter recoil, to also ensure that the stability condition is not violated.

It is largely a matter of choice whether to retard the counter recoiling parts throughout the counter recoil stroke or only towards the end of counter recoil. In most cases retardation is done throughout the stroke but increasingly towards the end.

With reference to Fig [3.6.1], taking moments about the point of contact of wheels with the ground:

$$F_{Bcr} \, Cos\varphi h + R_4 l = W(l - w)$$

Or:

$$R_4 = \frac{W(l - w) - F_{Bcr} Cos\phi h}{l}$$

For stability, the trails must not leave the ground i.e. R_4 must be equal to 0.

Hence for stability:

$W(l\text{-}w)$ is greater than or equal to $F_{Bcr} \, Cos\varphi h$

If a factor of safety s_{cr} is introduced:

$$F_{Bcr} = s_{cr} \frac{W(l - w)}{hCos\phi} \quad \text{[3.6.2]}$$

Energy of Counter Recoil

Mechanical Spring Recuperator

With reference to Fig 3.6.2 below, from similar triangles:

$$\frac{F_m}{F_{r(0)}} = \frac{\delta_0 + l_r}{\delta_0}, \text{ from which:}$$

$$F_m = F_{r(0)} \frac{\delta_0 + l_r}{\delta_0}$$

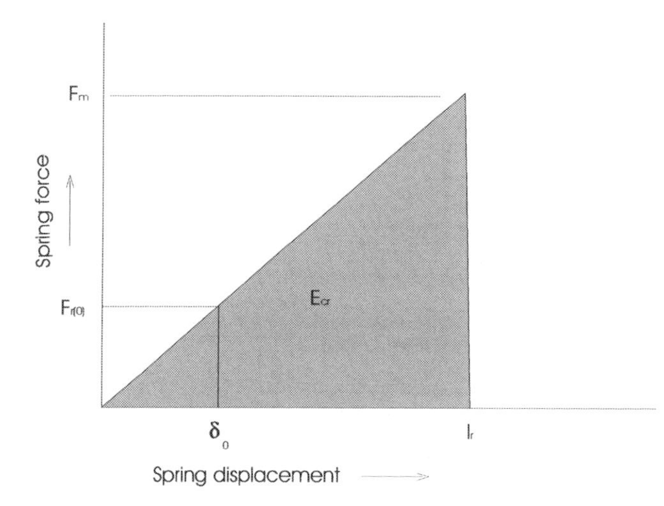

Fig 3.6.2: Spring force-displacement diagram of a mechanical spring

The energy required to return the gun to the firing position is stored in the recuperator. In the case of a mechanical spring recuperator, the total available counter recoil energy is given by:

$$E_{cr} = \frac{1}{2} F_m (\delta_0 + l_r), \text{ substituting for } F_m:$$

$$E_{cr} = \frac{1}{2}\frac{F_{r(0)}}{\delta_0}[\, l_r + \delta_0]^2 \quad \text{[3.6.3]}$$

$F_{r(0)}$: initial recuperator force
F_m: spring force at maximum compression
δ_0 : initial spring compression
l_r: recoil length

Some of this energy is utilized in overcoming the inescapable resistance to counter recoil i.e. the weight component acting against the direction of counter recoil, the sliding friction force and the gland friction. The energy of the counter recoiling parts available to return the gun to the firing position is finally:

$$E_{cr} = \frac{1}{2}\frac{F_{r(0)}}{\delta_0}[l_r + \delta_0]^2 - F_{r(0)}\, l_r \quad \text{[3.6.4]}$$

The energy available when the buffer is contacted is given by:

$$E_b = \frac{1}{2}\frac{F_{r(0)}}{\delta_0}[\, l_b + \delta_0]^2 - F_{r(0)}\, l_b \quad \text{[3.6.5]}$$

l_b: buffer stroke

Hydro Pneumatic Systems

In a hydro-pneumatic recuperator the polytropic compression of gases follows the law pV^n is a constant.

p: pressure of gas in recuperator at a volume V
p_1: gas pressure at beginning of recoil stroke, corresponding to the initial recuperator force
p_2: gas pressure at end of recoil stroke
V_1: volume of gas at beginning of recoil stroke
V_2: volume of gas at end of recoil stroke
n: polytropic exponent

From the gas laws: $pV^n = p_1 V_1^n = p_2 V_2^n$ ....[3.6.6]

Work done in compressing the gas is equal to the energy stored in the compressed gas of the recuperator:

$$E_{cr} = \int_{v_1}^{v_2} p\,dV = \int_{v_1}^{v_2} \frac{p_1 V_1^n}{V^n} dV = p_1 V_1^n \int_{v_1}^{v_2} V^{-n} dV$$

$$= p_1 V_1^n \left[\frac{V^{-n+1}}{-n+1} \right]_{v_1}^{v_2}$$

$$= \frac{1}{1-n} \left[p_1 V_1^n V_2^{1-n} - p_1 V_1^n V_1^{1-n} \right]$$

From equation [3.6.6]: $p_1 V_1^n = p_2 V_2^n$

$$E_{cr} = \frac{p_2 V_2 - p_1 V_1}{1-n} \quad \text{..} [3.6.7]$$

The energy available when the buffer is contacted can be computed knowing the buffer stroke and hence the volume.

$$E_b = \frac{p_b V_b - p_1 V_1}{n - 1} \qquad [3.6.8]$$

During counter recoil:

E_b: Energy of the recuperator when the buffer is contacted
p_b: pressure of the gas when buffer is contacted
A: area of the recuperator floating piston
$V_b = Al_b$: volume of the gas when buffer is contacted

Velocity of Counter Recoil

Assuming that no braking force is applied, the average velocity of counter recoil is given by:

$$v_{cr} = \sqrt{\frac{2E_r}{m_g}} \qquad [3.6.9]$$

m_g: mass of the counter recoiling parts

From the value of counter recoil velocity so calculated, it is possible to assess whether the counter recoil velocity is too high or too low. If too low for efficient run out, more energy is to be stored. If the velocity is too high, the fluid flow has to be restricted during the counter recoil stroke.

Sequence of Counter Recoil

At the end of recoil, a part of the piston rod is outside the recoil brake cylinder. Hence there is a vacuum in the recoil brake cylinder of volume equal to that of the volume of the piston rod outside the recoil brake cylinder. This partial vacuum allows unrestricted acceleration of the counter recoiling parts at the commencement of counter recoil. After the gun has counter recoiled some distance, this vacuum is taken up and oil will flow in the reverse direction, whether through the recoil brake piston orifices or other designated passages to assume its earlier position in the recoil brake cylinder. This restricted flow generates a retarding force during counter recoil. From the time this force is introduced onwards, the run out velocity will be fairly constant, as the counter recoil force of the recuperator will be balanced by the recoil brake retardation and also the friction forces. This velocity must be reduced smoothly to zero, for the counter recoil to end without impact. This requirement necessitates the introduction of an additional retardation to ensure zero velocity at the end of counter recoil. This additional retardation is the function of what is called the buffer. The buffer is usually a dashpot arrangement as illustrated below.

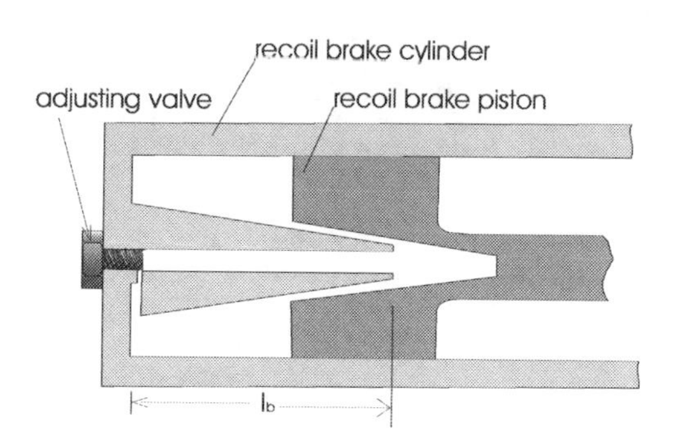

Fig 3.6.3: Stroke length of dashpot type buffer

Retarding Force of the Buffer

The magnitude of the buffer force is based on the energy of counter recoil and the buffer stroke length. The buffer force during counter recoil is:

$$F_b = \frac{E_b}{l_b} \quad \text{[3.6.9]}$$

E_b: energy of the recoiling parts when it encounters the buffer
l_b: buffer stroke length.

The control rod, which is tapered towards the piston, is attached to the recoil brake cylinder and the chamber is formed by a cavity of constant cross section in the piston. During run out when the control rod enters the chamber, liquid is trapped in the chamber and can only escape through the clearance between the control rod and the chamber walls. Due to the tapering of the control rod, this clearance between the rod and the chamber walls decreases, thereby increasing the resistance to counter recoil.

Control Valve

A control valve is incorporated which controls the flow of liquid through a central drilling in the control rod so that fine adjustment to counter recoil is possible. This may be necessary due to temperature variations, leakage of liquid or change in the viscosity of the liquid over a period of time.

Stages of Counter Recoil

The counter recoil may be seen in three distinct stages.

- During the first stage, which commences at the end of recoil, the gun is accelerated by a force equal to the difference between the recuperator force and the friction force. The recuperator force will decrease with displacement, in proportion to its spring stiffness in the case of mechanical spring recuperators or with the expansion of gas in the case of HP recuperators. The acceleration will commence from zero, attain a maximum value as the inertia and net friction forces are overcome and then fall in proportion to the recuperator force. During this stage the vacuum in the cylinder, caused by the withdrawal of some length of the piston rod, is taken up, pressure does not yet act on the liquid in the cylinder, and hence the recoil brake force is zero. The equation of motion of the counter recoiling parts is given by:

$$F_r - F_f - F_g = m_g \frac{dv_{cr}}{dt} \qquad [3.6.10]$$

F_r: recuperator force
F_f: friction force of slides
F_g: friction force of glands.

It follows that acceleration is positive and velocity increasing.

- During the second stage, the acceleration is kept at zero by the recoil brake resistance, which in the ideal condition is equal and opposite to the recuperator force less the friction force.

$$F_r - F_{Bcr} - F_f - F_g = 0 \qquad [3.6.11]$$

Here F_{Bcr} is the braking force introduced in the recoil brake by restricting the reverse flow of the liquid. Acceleration during this stage is zero and velocity constant.

- During the third and final stage, the retardation is increased by the introduction of the buffer retardation force F_b. Accordingly the velocity is brought to zero.

$$F_r - F_{Bcr} - F_f - F_g - F_b = m_g \frac{dv_{cr}}{dt} \qquad [3.6.12]$$

F_b: buffer force

Here $F_r < F_{Bcr} + F_f + F_g + F_b$ and decreasing, hence the right hand side of the equation becomes negative which implies retardation.

3.7: Recoil Brakes

Components of a Typical Recoil System

A recoil system consists basically of the following functional sub systems:

- The recoil brake, which may include the sub assembly, called the buffer.
- The recuperator.

The actual configuration differs from design to design but the functions of recoil braking, recuperation or counter recoil and buffing are performed interdependently by the sub systems of the recoil system.

An essential feature of recoil systems is effective sealing to prevent escape of oil and gas to the atmosphere as also intermingling of the two, where in contact.

Recoil Brake

This invariably consists of a hydraulic cylinder and piston assembly. Braking is achieved by restricting the flow of liquid from one side of the piston to the other. As the hydraulic braking force varies, the restriction may be achieved by varying the cross sectional area of orifices in the piston head itself. It may be achieved by means of orifices in the shape of variable depth grooves, cut into the cylinder inner wall surface or by a rod of varying cross section, called a control rod, moving inside the hollowed piston rod. In this case the varying gap between the aperture in the piston and the control rod serves as the orifice. The magnitude of this restricting force depends on the cross sectional area and the shape of the orifices through which the liquid flows as also the velocity of counter recoil. The energy generated by the restriction of movement of the liquid is dissipated as heat. Different types of recoil brakes are depicted in Fig 3.7.1 ahead.

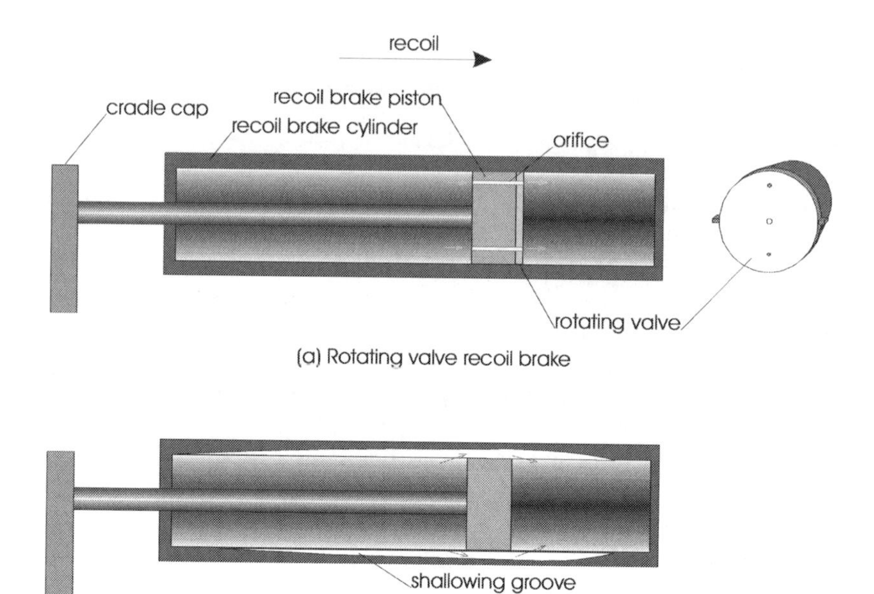

(a) Rotating valve recoil brake

(b) Shallowing groove recoil brake

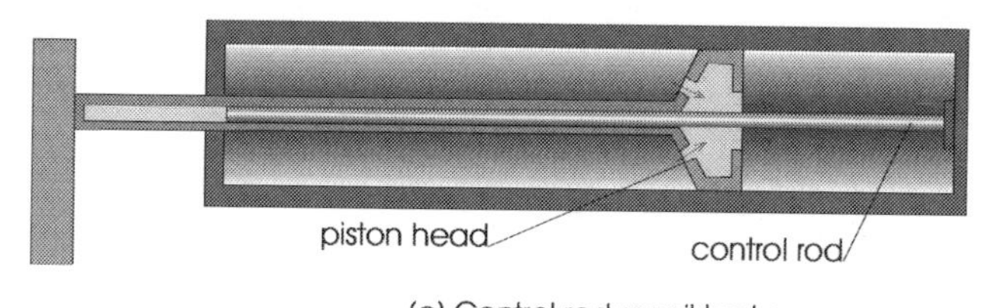

(c) Control rod recoil brake

Fig 3.7.1: Types of recoil brakes

Components of the Recoil Brake:

- Recoil cylinder: The recoil cylinder is made of rigid material to minimize the dilation due to the internal pressure, which would reduce the effectiveness of the seals. The thickness of the cylinder is related to the properties of the material used and is calculated based on the appropriate theory of failure of the material, with an adequate factor of safety.

- Recoil Piston Rod. The recoil piston rod is a tension member. One end is attached to the piston, the other end to the cradle. The rod may be threaded to the piston or integral with it. Thick and rigid rods are preferred to thin rods, which are likely to bend and result in damage to the seals, which will result in leakage of liquid from the system.

- Recoil Piston. The thickness of the piston depends on the space required for the packing. The piston area and hence the diameter is dictated by the peak pressure reached in the cylinder. This pressure is again limited by the efficiency of the sealing. With better packing materials, higher peak pressures are possible.

$$A_{B_p} = \frac{F_H}{p_m} \qquad [3.7.1]$$

A_{Bp}: effective piston area of the recoil brake
F_H: recoil brake hydraulic force
p_m: peak pressure in the recoil brake cylinder

The piston diameter is determined from:

$$A_{B_p} = \frac{\pi}{4}\left(d_{Bi}^2 - d_{Br}^2\right)$$

d_{Bi}: piston diameter or inner diameter of the cylinder
d_{Br}: piston rod diameter.

Materials for Recoil System Components

Except for the bearings, bushings and packings, all components of the recoil mechanism are made of steel. The bearings are of anti-friction metals like tin, antimony, copper and the bushings are of bronze. Moderate yield strength steels are used instead of high strength steels, as rigidity is of importance.

Theory of Hydraulic Recoil Brakes

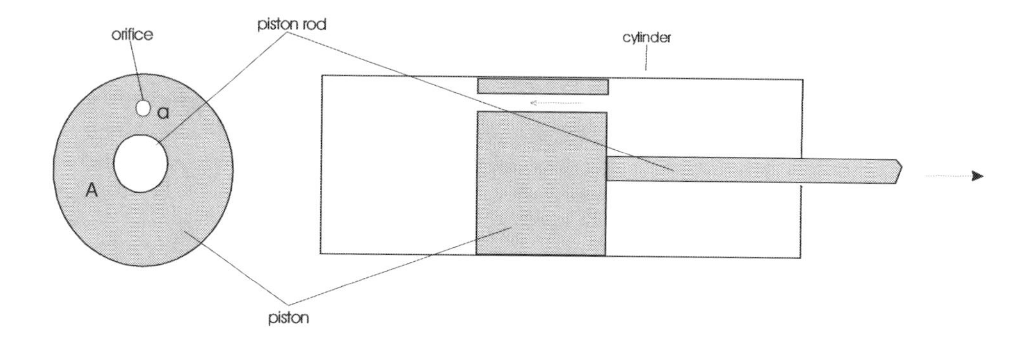

Fig 3.7.2: Flow of liquid through piston orifice

a: orifice area cross-section.
A_{Bp}: effective area of piston head
ρ: mass density of liquid in the cylinder
v_g: velocity of recoil
u: velocity of liquid flowing through the orifice, relative to the piston head.

Two arrangements of a hydraulic recoil brake are possible. In the first, the recoil brake cylinder is fixed to a bracket on the cradle, the piston moving with its rod attached to the recoiling parts. The second arrangement has the piston rod fixed to the extension of the cradle, while the cylinder moves with the recoiling parts.

Recoil Brake with Recoil Cylinder Fixed, Piston Moving

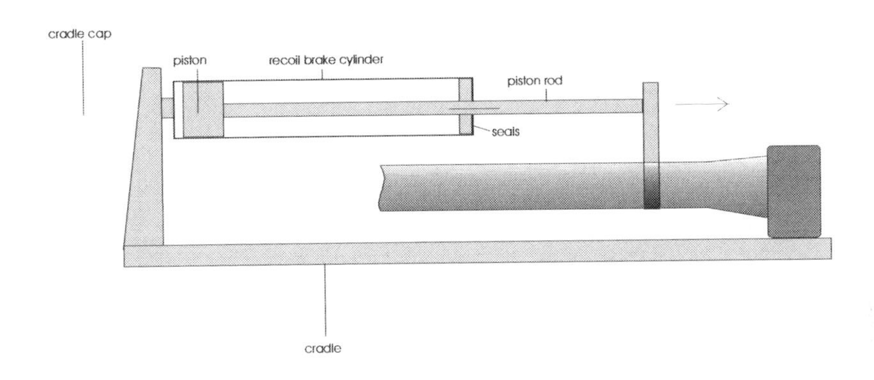

Fig 3.7.3: Fixed cylinder moving piston type hydraulic brake

Assuming that the liquid is ideal, the volume of liquid forced through the orifice in unit time is:

$$V = v_g A_{B_p} \qquad [3.7.2]$$

In terms of the liquid velocity and area of the orifice, the volume of liquid that flows through the orifice in the same time is:

$$V = ua \qquad [3.7.3]$$

Hence from Equations [3.7.2] and [3.7.3]:

$$u = \frac{v_g A_{B_p}}{a} \qquad [3.7.4]$$

Velocity of liquid emerging from the orifice is:

$$u\text{-}v_g = \frac{(A_{B_p} - a)v_g}{a}$$

This implies that in unit time a mass of liquid $v_g A_{B_p} \rho$ changes its velocity from zero to:

$$\frac{(A_{B_p} - a)v_g}{a}$$

The kinetic energy imparted to the liquid in unit time is:

$$\frac{1}{2} v_g A_{B_p} \rho \left[\frac{(A_{B_p} - a)v_g}{a} \right]^2 \qquad [3.7.5]$$

Consider three possible conditions:

- Firstly, if $a = A_{B_p}$, there is no change in the velocity of the liquid. Hence no kinetic energy is imparted to it. In other words, the oil flows freely from one side of the piston to the other and there is no resistance to the motion of the piston.

- Secondly if $a = 0$, the liquid being nearly incompressible, the mechanism of piston and cylinder behaves as one rigid body and the entire recoil force acts on the high-pressure side of the piston, and from here via the cradle and the trunnions to the supporting structure.

- Thirdly if $a > A_{B_p}$, a being greater than zero, a retarding force acts on the high-pressure side of the piston. The work done by this force being equal to the kinetic energy imparted to the liquid.

If F_H is the hydraulic force retarding the movement of the piston in unit time, the work done by this force is $F_H v_g$.

$$F_H v_g = \frac{\rho v_g^3 A_{B_p} \left(A_{B_p} - a\right)^2}{2a^2} \text{ i.e.}$$

$$F_H = \frac{\rho v_g^2 A_{B_p} \left(A_{B_p} - a\right)^2}{2a^2} \quad \text{[3.7.6]}$$

Orifice Cross Section

As a is much smaller than A_{Bp}, it can be approximated without serious error that:

$$F_H = \frac{\rho v_g^2 A_{B_p}^{\ 3}}{2a^2}$$

Rearranging:

$$a = v_g \sqrt{\frac{A_{B_p}^3 \rho}{2F_H}} \quad \text{[3.7.7]}$$

It can be seen that at any instant during recoil, the orifice cross section is a function of the velocity of the recoiling parts and the retarding force.

Recoil Brake with Recoil Brake Piston Fixed Cylinder Moving

The velocity of the liquid relative to the piston is $u = \frac{A_{B_p} v_g}{a}$.

As the piston is fixed, this is the actual velocity of the liquid.

In unit time, the liquid changes its velocity from v_g to $\frac{A_{B_p} v_g}{a}$.

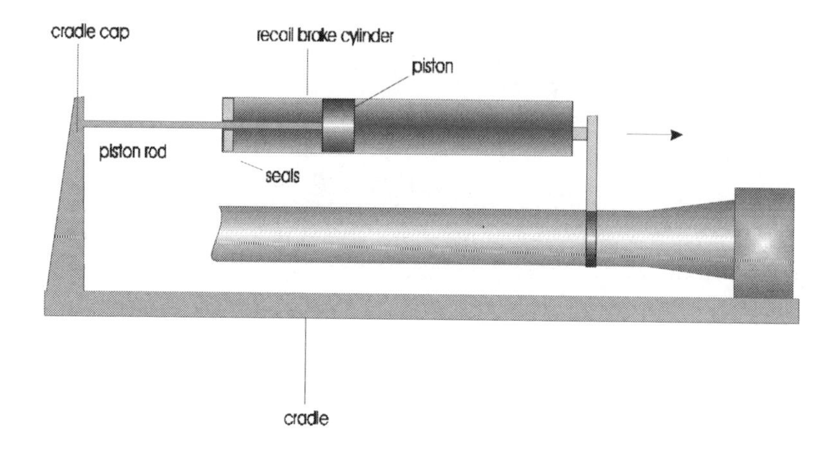

Fig 3.7.4: Fixed piston-moving cylinder type recoil brake

The change in kinetic energy is:

$$\frac{1}{2}v_g A_{B_p}\rho\left[\left(\frac{A_{B_p}v_g}{a}\right)^2 - v_g^2\right]$$

Therefore:

$$F_H v_g = \frac{1}{2}v_g A_{B_p}\rho\left[\left(\frac{A_{B_p}v_g}{a}\right)^2 - v_g{}^2\right]$$

Or:

$$F_H = \frac{1}{2a^2}v_g^2 A_{B_p}\rho\left[A_{B_p}^2 - a^2\right] \qquad [3.7.8]$$

This is identical to the case of fixed piston moving cylinder. Practically, the actual velocity of a non-ideal liquid is influenced by its viscosity and the profile of the orifice shape through which it flows. A factor of around 2/3 is applied to the orifice cross sectional area a, which takes into account the actual properties of the liquid and the profile of the orifice.

Maximum Pressure in the Cylinder

The maximum pressure in the cylinder on the high-pressure side is:

$$p_m = \frac{F_H}{A_{B_p} - a} \quad \text{[3.7.9]}$$

This maximum pressure influences the strength of the cylinder material and its thickness. The strength of the packings may be a limiting factor on the maximum pressure, which can be applied in the cylinder. The main problem of recoil brake design is the variation of orifice area in accordance with the instantaneous value of the braking force and recoil velocities at different instants of recoil.

3.8: Recuperators

Recuperation

Some of the energy of recoil is stored in a spring or by means of compressing a gas. The stored energy is used to return the gun to the firing position. The stored energy must be sufficient to return the gun to this position at the maximum elevation i.e. when the weight component of the gun is the maximum, as also against the frictional resistance of the seals of the recoil system. The initial recuperator thrust is sufficient to hold the gun forward at the maximum elevation. As the recoil progresses, this thrust increases and is finally in excess of what is actually required. The theory of two types of recuperators will be studied here; the mechanical spring type and the pneumatic type.

Spring Type Recuperator

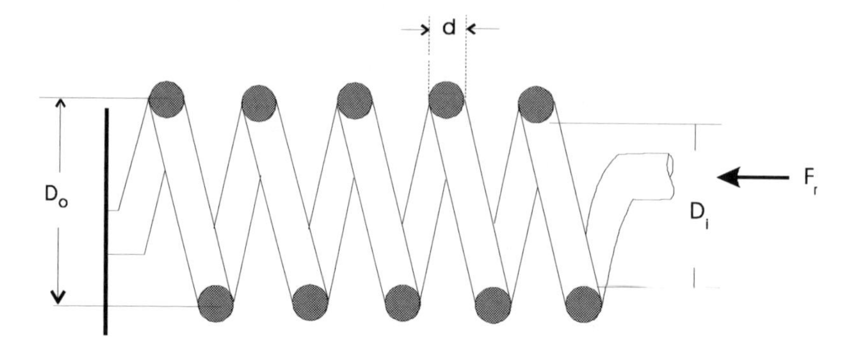

Fig 3.8.1: Helical spring

F_r: axial load on the spring, in this case the recuperator force
k: spring stiffness
D: mean coil diameter
D_i and D_0: inner and outer coil diameters respectively
d: diameter of the spring wire
$c=D/d$: spring index
n: number of effective coils
l: effective solid length
δ : deflection under load F_r
G: Modulus of Rigidity of the material of the spring

From the torsion of a circular shaft, shear stress induced in the wire due to torque $\frac{F_r D}{2}$ is given by:

$$\tau = \frac{8F_r D}{\pi d^3} \qquad [3.8.1]$$

From which:

$$F_r = \frac{\tau\pi d^3}{8D} \quad \text{.......} [3.8.2]$$

Also twist of one end relative to the other is:

$$\theta = \frac{16F_r nD^3}{Gd^4} \quad \text{.......} [3.8.3]$$

Hence the deflection of one end with respect to the other is:

$$\delta = \frac{8F_r nD^3}{Gd^4} \quad \text{.......} [3.8.4]$$

Evidently the deflection of the spring is directly proportional to the load. It follows that the load deflection curve of a closed coil spring is a straight line. The work done on the spring is the area under the load deflection curve and is equal to the energy U stored in the spring.

$$U = \frac{1}{2}F_r\delta$$

Or:

$$U = \frac{1}{2}F_r\frac{8F_r nD^3}{Gd^4}$$

$$U = \frac{4nD^3}{Gd^4}\left(\frac{\pi d^3\tau}{8D}\right)^2$$

$$U = \frac{\tau^2}{4G}\left(\pi Dn\frac{\pi d^2}{4}\right)$$

In other words:

$$U = \frac{\tau^2}{4G}.volume\ of\ wire \qquad [3.8.5]$$

Spring Type Recuperator

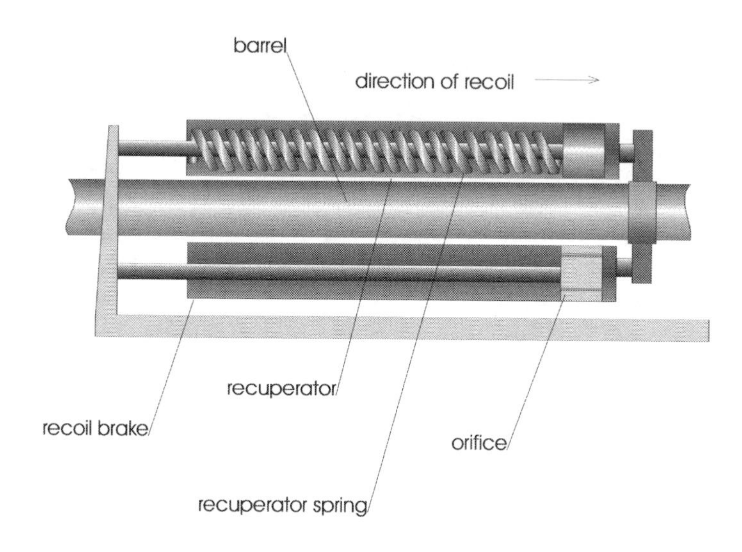

Fig 3.8.2: Spring type recuperator

If the recuperator spring shown in Fig: 3.8.2 is compressed under a load F_r to deflect by δ, then energy stored in the spring is $\frac{1}{2}F_r\delta$

If the spring under an initial load $F_{r(0)}$ and a deflection δ_0 is further compressed by a distance l_r to a deflection $l_r + \delta_0$ then:

$k = \frac{F_{r(o)}}{\delta_0}$ and the final load is $\frac{F_{r(0)}}{\delta_0}[l_r + \delta_0]$.

The energy stored at full compression is:

$$E = \frac{1}{2}\frac{F_{r(0)}}{\delta_0}[l_r + \delta_0]^2$$

$$= \frac{1}{2}F_{r(0)}\left[4l_r + \left(\frac{l_r}{\delta_0}\right)^2 - 2l_r + \sqrt{\delta_0}^{\,2}\right]$$

$$= \frac{1}{2}F_{r(0)}\left[4l_r + \left(\frac{l_r}{\delta_0}\right)^2 - 2r\frac{\sqrt{\delta_0}}{\sqrt{\delta_0}} + \sqrt{\delta_0}^{\,2}\right]$$

Or:

$$E = \frac{F_{r(0)}}{2}\left[4l_r + \left(\frac{l_r}{\sqrt{\delta_0}} - \sqrt{\delta_0}\right)^2\right] \text{.........} [3.8.6]$$

If $F_{r(0)}$ and l_r are constant then the energy is at its minimum when $l_r = \delta_0$

Or:

$$E = 2F_{r(0)}\delta_0$$

It follows that if the condition for minimum energy stored is satisfied, the recuperator force at the end of recoil is:

$$F_r = \frac{2F_{r(0)}}{\delta_0}\delta_0$$

Or:

$$F_r = 2F_{r(0)} \text{.........} [3.8.7]$$

In a recuperator, the initial recuperator force is fixed. The recoil length is also fixed with due consideration to the stability condition or space restrictions. It is seen that the energy stored is a minimum when the initial compression of the springs is equal to the recoil length or in other words when the final recuperator force equals twice the initial recuperator force. As $F_{r(0)}$, the initial recuperator force and the length of recoil are both known, the spring stiffness required can then be found . The physical dimensions D, n and d and the material specify a helical spring. Once the permissible value of stress is fixed, the stiffness can be calculated according to the criterion of minimum energy.

Pneumatic Recuperators

Isotropic Compression of Air

When a gas is compressed adiabatically, the law:

pv^{γ} = a constant, holds good.

$\gamma = c_p / c_v = 1.41$ for dry air.

However during gas compression in a recuperator, the process cannot be taken to be adiabatic as some heat is lost due to the work done in compressing the gas.

By applying a value of an index $m = 1.22$ instead of $\gamma = 1.41$, the compensation is effected. Hence for recuperators the acceptable expression is:

pv^{m} = a constant

Compression Ratio

If p_0 and v_0 are the initial pressure and volume of the gas:

$$p = p_0\left(\frac{v_0}{v}\right)^m = p_0 c_r^{\ m} \quad \text{[3.8.8]}$$

c_r: compression ratio.

The negative work done in compressing a gas from initial volume v to final volume $v\text{-}dv$ is:

$$dW = -pdv$$

Since:

$$v = \frac{v_0}{c_r} \Rightarrow dv = -\frac{v_0}{c_r^2}dr \Rightarrow dW = p\frac{v_0 dr}{c_r^2}$$

Substituting for p from Equation [3.8.8]

$$dW = p_0 c_r^m \frac{v_0 dr}{c_r^m} = p_0 v_0 c_r^{m-2} dr$$

The work done in compressing the gas from v_0 to v is;

$$\int dW = \int_{v_0}^{v} p_0 v_0 c_r^{m-2} dr$$

Or:

$$\int dW = \int_{1}^{c_r} p_0 v_0 c_r^{m-2} dr .$$

Hence:

$$W = \int_1^{c_r} p_0 c_r^m \frac{v_0}{c_r^2} dr = \frac{p_0 v_0}{m-1} {}_1^{c_r}\left[c_r^{m-1}\right]$$

$$W = \frac{p_0 v_0}{m-1}\left[c_r^{m-1} - 1\right] \qquad [3.8.9]$$

Types of Pneumatic Recuperators

Recuperators in which energy is stored by compression of a gas are called pneumatic; they may be of the dry air type or the liquid and gas or hydro pneumatic type.

Dry Air Recuperators

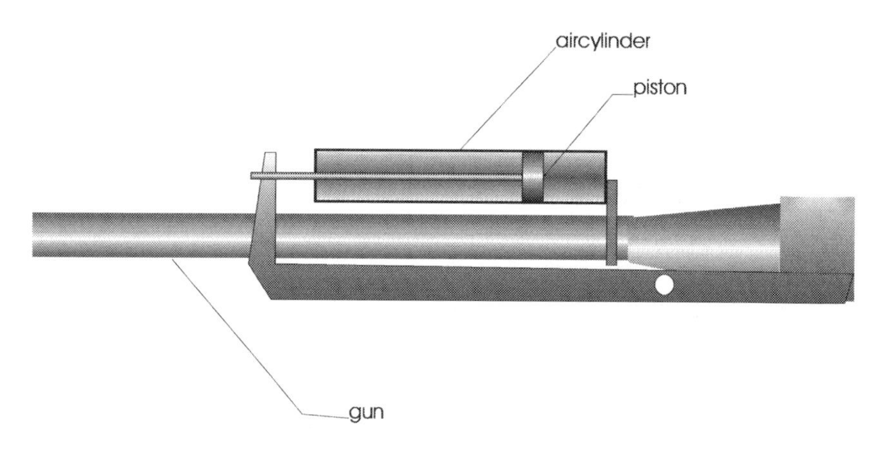

Fig 3.8.3: Dry air recuperator

A_R: effective area of the recuperator piston
l_0: initial length of the air column
x_g: distance recoiled at any instant
p_0: initial recuperator pressure

In the dry air recuperator, gas under pressure is contained in a cylinder closed by a piston. During recoil, the cylinder and piston move relatively, further compressing the gas.

Initial volume of the air: $V_0 = A_R\, l_0$

Final volume of the air: $V = A_R(l_0 - x_g)$

The compression ratio $c_r = \dfrac{l_0}{l_0 - x_g}$

The recuperator force F_r is given by:

$$F_r = A_R p = A_R p_o c_r^m = F_{r(0)}\left[\frac{l_0}{l_0 - x_g}\right]^m \quad \text{[3.8.10]}$$

The work done in compressing the gas from l_0 to $l_0 - x_g$ is

$$W = \frac{\frac{F_{r(0)}}{A} A l_0}{m-1}\left[c_r^{m-1} - 1\right]$$

$$W = \frac{F_{r(0)} l_0}{m-1}\left[\left(\frac{l_0}{l_0 - x_g}\right)^{m-1} - 1\right] \quad \text{[3.8.11]}$$

Hydro Pneumatic Recuperators

This arrangement consists of a twin cylinder filled with oil, interconnected by means of an oil passage. One cylinder serves as the recuperator cylinder and the other consists of a floating piston, which separates an air cylinder from the oil component. On recoil, oil is forced from the recuperator cylinder to the oil side of the floating piston. The incompressibility of oil forces the floating piston to further compress the air, already at a pressure sufficient to maintain the gun in the run up position. At the end of recoil, the compressed air expands driving the oil back into the recuperator cylinder thereby forcing the gun to run out.

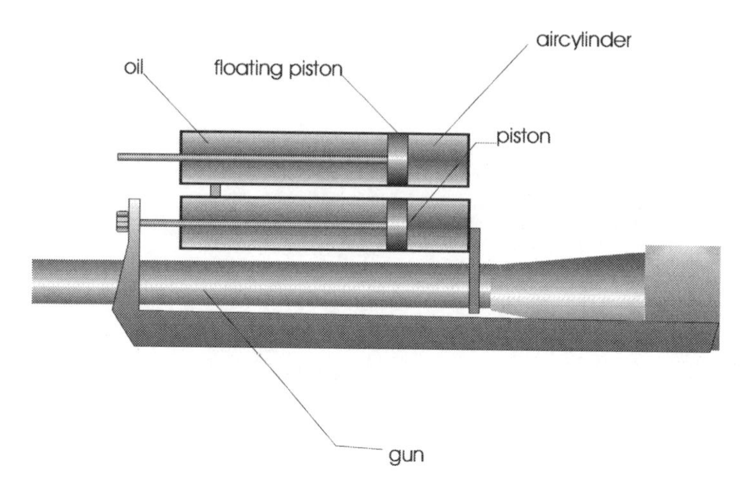

Fig 3.8.4: Hydro-pneumatic recuperator

Theory of the Hydro-pneumatic (HP) Recuperator

A_{F1}: area of the floating piston on the air side.
A_F: effective area of the floating piston on the liquid side.
A_R: effective area of the recuperator piston in contact with liquid
l_0: initial length of the air column

When the recoil distance is x_g, the volume of oil forced into the HP cylinder is $A_R x_g$. Hence the travel of the floating piston is $x_F = \frac{A_R x_g}{A_F}$. If the air pressure is p then force on the floating piston is pA_{F1}. Pressure in the liquid is hence $\frac{pA_{F1}}{A_F}$ and force on the recuperator piston is:

$$F_r = p\frac{A_{F1}A_R}{A_F} \qquad [3.8.12]$$

If the initial air pressure is p_0, it follows that the initial recuperator force is:

$$F_{r(0)} = p_0 \frac{A_{F1}A_R}{A_F}$$

We have $pV^m = p_0V_0^m$, from which $p = p_0c_r^m$

c_r^m : compression ratio

Equation [3.8.12] becomes:

$$\therefore F_r = F_{r(0)}c_r^m$$

$$F_r = F_{r(0)}\left[\frac{l_0}{l_0 - x_F}\right]^m \quad [3.8.13]$$

Substituting for x_F:

$$F_r = F_{r(0)}\left[\frac{1}{1 - \frac{A_R x_g}{A_F l_0}}\right]^m$$

If the maximum recoil displacement is l_r, the work done in compressing the air is:

$$W = \int_0^{l_r} F_r dx_g = F_{r(0)} \int_0^{l_r} \left(1 - \frac{A_R x_g}{A_F l_0}\right)^{-m} dx_g$$

Substituting:

$$u = 1 - \frac{A_R x_g}{A_F l_0}, \quad \frac{du}{dx_g} = -\frac{A_R}{A_F l_0} \text{ and } dx_g = -\frac{A_F l_0}{A_R} du$$

At the start of recoil:

$x_g = 0,\ u = 1$

At the end of recoil:

$x_g = l_r$ and $u = 1 - \frac{A_R l_r}{A_F l_0}$

The work done in compressing the air can be written as:

$$W = -F_{r(0)} \frac{A_F l_0}{A_R} \int_1^{1-\frac{A_R l_r}{A_F l_0}} u^{-m} du$$

Or:

$$W = -F_{r(0)} \frac{A_F l_0}{A_R} \left[\frac{u^{1-m}}{1-m} \right]_1^{1-\frac{A_R l_R}{A_F l_0}}$$

Finally:

$$W = \frac{F_{r(0)}}{m-1} \frac{A_F l_0}{A_R} \left[\left(1 - \frac{A_R l_R}{A_F l_0} \right)^{1-m} - 1 \right] \quad [3.8.14]$$

Intensification

As the area on the air side of the floating piston A_{F1} is greater than the area on the oil side A_F, the pressure on the oil side is greater than that on the air side, this ensures any leakage is only from the oil side to the air side and not vice versa. This is called intensification.

3.9: Seals & Sealing

Seals and Sealing

Seals prevent leakage past moving parts like pistons and piston rods. The seals are forced firmly against the moving surfaces by the fluid pressure and by the use suitably inserted springs. A sectionized arrangement of a seal is shown ahead.

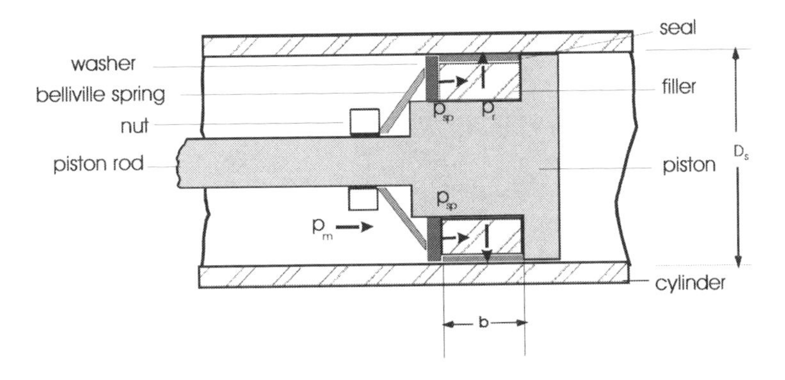

Fig 3.9.1: Sectionized recoil system seal

Because under high pressures, the sealing material behaves hydrostatically, the axial pressure is nearly equal to the radial pressure, which is necessary for sealing. The ratio of the axial pressure to the radial pressure is a property of the sealing material and is called the pressure factor K_p.

$$K_p = \frac{p_a}{p_r} \qquad [3.9.1]$$

K_p: pressure factor
p_a: axial pressure
p_r: radial pressure

The Pressure Factor is a constant for a given filler material and is akin to Poisson's Ratio. To ensure positive sealing, the radial pressure must be greater than the maximum axial fluid pressure. This is made possible by the pressure applied by suitably placed springs.

The ratio of radial pressure to the maximum fluid pressure is known as the leakage factor υ.

$$\upsilon = \frac{p_r}{p_m} \qquad [3.9.2]$$

ν: leakage factor
p_m: maximum fluid pressure

The leakage factor generally has a value of 1. When a small amount of leakage is desirable for lubrication purposes, the leakage factor is designed to be less than 1.

The radial pressure exerted by the seals expressed in terms of fluid pressure is:

$$p_r = K_p\left(p_{sp} + p_m\right) = \upsilon p_m \qquad [3.9.3]$$

p_{sp}: axial pressure in packing produced by the spring.

Solving for p_{sp}:

$$p_{sp} = \frac{\nu - K_p}{K_p} p_m \qquad [3.9.4]$$

The spring pressure being known, it is now possible to find the gland or packing friction. The total axial pressure on the seal equals the spring pressure plus the fluid pressure:

Axial pressure on the seal:

$$p_a = p_{sp} + p_{xg}$$

Radial pressure on the seal:

$$p_r = K_p p_a$$

Contact area of the seal on the cylinder wall:

$$A_s = \pi D_s b$$

D_s: outer diameter of seal
b: width of seal
p_{xg}: fluid pressure on seal at any recoil displacement x_g

The frictional force of a seal assembly

$$F_s = \mu A_s p_a \qquad [3.9.6]$$

If the frictional force of the seals in the recuperator is F_{Rs} and that in the recoil brake is F_{Bs}, Then the total frictional force due to the seals is:

$$F_s = F_{Rs} + F_{Bs} \qquad [3.9.7]$$

Rubber sealing filler is normally used; the liner or seal ring in contact with the cylinder is of leather. The extrusion of the leather between piston rings and cylinder is prevented by the use of silver rings. Teflon and aluminum alloys are currently used in place of leather and silver respectively.

Belleville Springs

These springs are commonly used to augment the packing pressure. Belleville springs are selected because they provide high loads at small deflections in compact spaces, such as gun recoil mechanism cylinders. In their zero spring rate i.e. constant load form, they are commonly used to load seals, which need a constant load over a small deflection. However they are sensitive to small changes in dimensions, hence manufacturing variations are not acceptable.

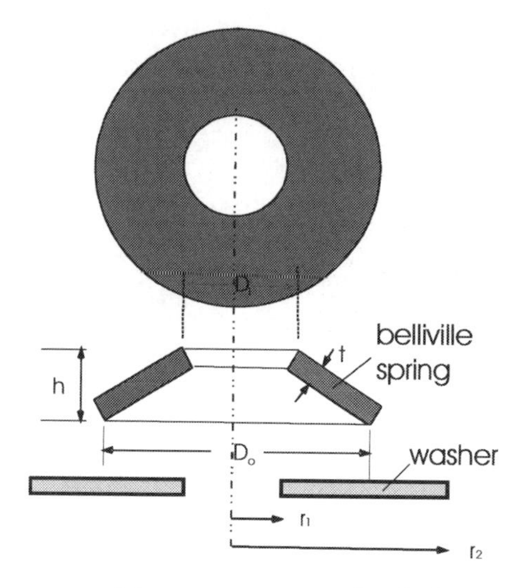

Fig 3.9.2: Section through a Belleville spring

The spring load required of the Belliville spring is:

$$F_{sp} = \pi\left(r_2^2 - r_1^2\right)p_{sp}$$

p_{sp}: packing pressure required of the spring
r_1 and r_2: the inner and outer radius of the seal ring washer respectively.

Values of Coefficient of Friction of Common Sealing Materials

For leather: $\mu = 0.05$
For silver: $\mu = 0.09$.

Pressure Factor for Rubber Filler

For rubber filler:

$K_p = 0.73$.

4

Muzzle Brakes

4.1: Principle and Purpose of a Muzzle Brake

Purpose of a Muzzle brake

The purpose of a muzzle brake is to reduce the backward momentum of the recoiling parts by deflecting some of the gases which emerge after shot ejection onto a baffle plate. The baffle plate is rigidly attached to the muzzle end of the barrel and may be considered a part of the barrel itself. If the rate of change of velocity of the gases is high, the thrust on the baffle plate will be greater. This is achieved by permitting maximum expansion of the gases before they strike the baffle plate and deflection of the gases through the maximum practicable angle after impact with the baffle plate. The gases attain maximum velocity if allowed to expand to the least pressure, which in this case, is atmospheric. The thrust on the baffle plate also depends on the mass flow rate of the gas impinging on it. Therefore the greater the percentage of mass of the gases diverted onto the baffle, the greater the thrust.

Energy of Recoil due to After Effect

The period from the instant of projectile exit, from the muzzle, when some of the gas following the projectile, and the projectile are at the same velocity, to the instant when the pressure falls to ambient, is called the after effect period. It is possible to quantify the influence of the gas pressure, on the recoil energy during this period using already established criteria.

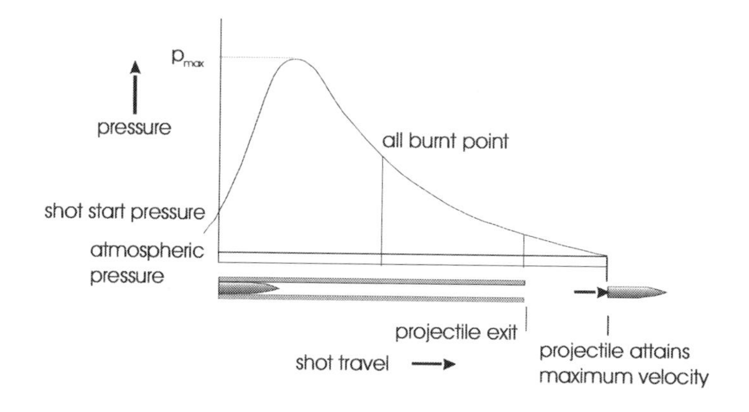

Fig 4.1.1: After effect period

E_g: energy of recoiling parts
v_p: projectile velocity
v_g: velocity of recoiling parts
m_g: mass of recoiling parts
m_p: mass of projectile
m_c: mass of charge
K: Krupps constant

Subscripts indicate conditions at:

(e): projectile exit
(a): end of after effect period

The increase in projectile velocity from projectile exit to the end of gas after effect being small, $v_{p(e)} = v_{p(a)}$ can be approximated. Hence the energy of the recoiling parts due to the after effect is given by:

$$E_{g(a)} - E_{g(e)} = \frac{1}{2} m_g \left(v_{g(a)}^2 - v_{g(e)}^2 \right)$$

Applying Equations [3.3.4] and [3.3.5]:

$$E_{g(a)} - E_{g(e)} = \frac{1}{2} \frac{v_{p(a)}^2}{m_g} \left[\left(m_p + K m_c \right)^2 - \left(m_p + \frac{1}{2} m_c \right)^2 \right]$$

$$= \frac{1}{2} \frac{v_{p(a)}^2 m_c}{m_g} \left[2K m_p + K^2 m_c - m_p - \frac{m_c}{4} \right]$$

$$= \frac{1}{2} \frac{v_{p(a)}^2 m_c}{m_g} \left[m_p (2K - 1) + m_c \left\{ \left(k - \frac{1}{2} \right) \left(k + \frac{1}{2} \right) \right\} \right]$$

$$= \frac{1}{2} \frac{v_{p(a)}^2 m_c}{m_g} \left[2m_p + m_c \left(K + \frac{1}{2} \right) \right] \left(K - \frac{1}{2} \right)$$

The ratio of the energy of the recoiling parts, due to after effect, to the energy of the recoiling parts at projectile exit is given by:

$$\frac{E_{g(a)} - E_{g(e)}}{E_{g(a)}} = \frac{m_c \left[2m_p + m_c \left(K + \frac{1}{2} \right) \right] \left(K - \frac{1}{2} \right)}{\left(m_p + K m_c \right)^2} \quad \text{.......... [4.1.1]}$$

Substituting for practical values of projectile weight and charge weight and taking the mean value of Krupp's constant which lies between 1.5 and 2.5, the ratio of the percentage of the energy due to after effect of the propellant gases, to the maximum energy of recoil, is illustrated with the help an example below.

Example 4.1.1

Calculate the percentage of energy due to the after effect of propellant gases, of the recoiling mass of a 76 mm gun, given that the projectile mass is 6.2 kg and the charge mass is 1.08 kg.

Solution: Example 4.1.1

$$\frac{E_{g(a)} - E_{g(e)}}{E_{g(a)}} = \frac{1.08[2.6.2 + 1.08(2 + 0.5)](2 - 0.5)}{(6.2 + 2.1.08)^2}$$

$$\frac{E_{g(a)} - E_{g(e)}}{E_{g(a)}} = 35\%$$

From the example above, it is clear that the after effect of the propellant gases increases the energy of recoil substantially. If this effect eliminated, the recoil energy will reduce considerably. Further if the gas velocity during the after effect period is made to strike on some protruding surface of the barrel; it would create a braking force applied in a direction opposite to that of recoil, further reducing the energy of recoil.

Practical Difficulties in Obtaining Ideal Conditions

The ideal situation is one in which all the gases exiting the barrel, are made to attain maximum velocity, strike a surface, called a baffle, and reverse their direction through 180°. The maximum velocity of the gases is attained when the gas pressure falls to the minimum, in this case ambient pressure. However, some practical problems preclude such an ideal situation. At best a compromise may be arrived at.

- In order to expand the gases to atmospheric pressure a large divergence of around 25 is required. The term divergence will be explained shortly.

- Some of the gases escape behind the projectile through the projectile passage, thereby reducing the amount of the gas striking the baffle.
- Due to risk of damage to the gun and likelihood of injury to the crew, it is not advisable to reverse the gas flow by 180 degrees.
- The gas pressure varies from instant to instant; hence a design of a muzzle brake taking the mean of the conditions over the overall period offers the best practical solution.

4.2: Gas Flow, Speed-up Factor and Divergence

Flow of Gas through a Gun Barrel

Consider the chamber of the gun as a reservoir, the muzzle as the throat, and the muzzle brake as the diverging portion of a converging diverging nozzle. Some simplifying assumptions are also necessary in order to proceed with the analysis of an otherwise complicated phenomena. These are enumerated below:

- A state of steady flow, or a succession of states of steady flow exist i.e. the mass rate of flow is constant.
- The process is adiabatic. This means that the gas neither receives heat from nor emits heat to it surroundings, this includes heat arising out of friction. Flow work is done at the cost of internal energy and the change in internal energy is equal to the flow work done by the gas.
- The effect of co volume is ignored.
- The ratio of specific heats of the gas is constant.
- Gas is presumed to fill the tube completely. This assumption is acceptable if the angle of divergence is less than around 30 degrees.
- The theory is based on one-dimensional compressible flow.
- Velocity of the gases in the chamber is negligible.

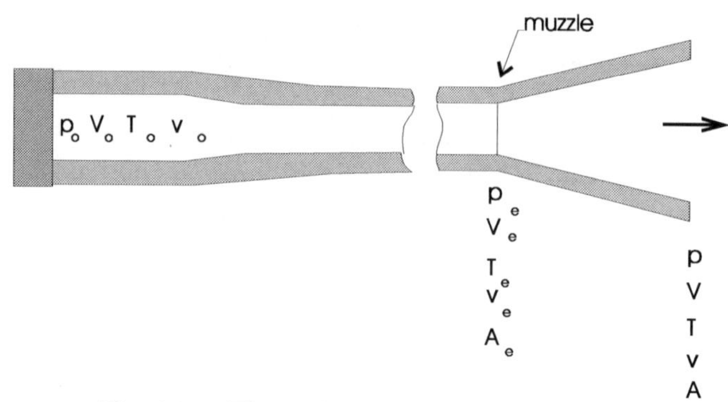

Fig 4.2.1: Flow of gas through the muzzle

p: pressure at a point
V: specific volume i.e. volume per unit mass
T: absolute temperature
v: gas linear velocity
A: cross sectional area of bore

Subscripts denote conditions at:

(0): chamber
(e): projectile exit
(a): end of after effect

From the gas laws:

$pV = RT$ and .... [4.2.1]

$pV^{\gamma} = p_{(o)}V_{(o)}^{\gamma}$ .... [4.2.2]

γ : ratio of specific heats of the gas.

From Bernoulli's theorem: Heat transferred to a fluid + Flow work + Work done on the fluid = Change in internal energy of the fluid + Change in its kinetic energy + change in its potential energy.

$$Q - \int_{p_0}^{p} V dp + work = u - u_{(0)} + \frac{v^2 - v_{(0)}^2}{2} + z - z_{(0)}$$

Q: heat transferred to the gas per unit mass
p: pressure.
V: specific volume
work: mechanical work done per unit mass
u: internal energy of the fluid per unit mass
z: elevation above the datum

Flow work: work done in pushing unit mass of the fluid across a section of the channel of area A. Force = $p.A$ and distance moved = V/A. Hence flow work done = pV.

Since the process is adiabatic, $Q = 0$, $work = 0$ and $z - z_{(0)} = 0$ and flow work is done at the expense of internal energy.

$$-\int_{p_0}^{p} V dp = \frac{v^2 - v_{(0)}^2}{2}$$

Writing $\frac{p}{p_{(0)}} = x$ and as in this case pV^{γ} = constant

$$x = \left(\frac{V_{(0)}}{V}\right)^{\gamma}$$

Or:

$$V = V_{(0)} x^{\frac{-1}{\gamma}} \qquad [4.2.3]$$

Substituting for V from Equation [4.2.3] and integrating between the limits $p = p_{(0)}$ and p

$$v^2 - v_{(0)}^2 = \frac{2p_{(0)}V_{(0)}\gamma}{\gamma - 1}\left[1 - x^{\frac{\gamma-1}{\gamma}}\right]$$

Writing:

$$\frac{\gamma - 1}{\gamma} = k$$

$$v^2 - v_{(0)}^2 = \frac{2p_{(0)}V_{(0)}}{k}\left(1 - x^k\right)$$

The velocity of the gases at the chamber may be safely assumed to be zero. Hence the equation immediately above reduces to:

$$v^2 = \frac{2p_{(0)}V_{(0)}}{k}\left(1 - x^k\right) \quad \text{[4.2.4]}$$

Based on the assumption that the flow is steady, the mass rate of flow $\dot{m}$ across a cross section A is:

$$\dot{m} = \frac{Av}{V}$$

Hence the mass rate of flow per unit cross section is:

$$\dot{m}\frac{1}{A} = \frac{v}{V}$$

Squaring and substituting for v from equation [4.2.4]

$$\left(\dot{m}\frac{1}{A}\right)^2 = \frac{1}{V^2}\left[\frac{2p_{(0)}V_{(0)}}{k}\left(1 - x^k\right)\right]$$

$\because \frac{1}{\gamma} = 1 - k$ and $V = V_{(0)} x^{\frac{-1}{\gamma}}$

$$\left(\dot{m}\frac{1}{A}\right)^2 = \frac{x^{2(1-k)}}{V_{(0)}}\left[\frac{2p_0}{k}\left(1 - x^k\right)\right] \qquad \text{[4.2.5]}$$

At the muzzle, the mass flow rate per unit cross section is maximum, hence the derivative of the right hand side of Equation [4.2.5] with respect to $x = 0$.

Putting:

$$z = \frac{x^{2(1-k)}}{V_{(0)}} \text{ and } y = \left[\frac{2p_0}{k}\left(1 - x^k\right)\right]$$

Differentiating logarithmically:

$$\frac{1}{z}\frac{dz}{dx} + \frac{1}{y}\frac{dy}{dx} = 0, \text{ or}$$

$$\frac{V_{(0)}}{x^{2(1-k)}}\frac{2(1-k)x^{2(1-k)-1}}{V_{(0)}} - \frac{2kp_{(0)}x^{k-1}}{2p_{(0)}\left(1 - x^k\right)} = 0$$

As this condition is true only at the muzzle, denoting values at the muzzle by *(e)*:

$$\frac{2(1-k)}{x_{(e)}} - \frac{kx_{(e)}^{k-1}}{\left(1 - x_{(e)}^k\right)} = 0$$

From which:

$$x_{(e)}^k = \frac{2(1-k)}{2-k} \qquad \text{[4.2.6]}$$

From Equations [4.2.4] & [4.2.6], velocity at the muzzle is given by:

$$v_{(e)}^2 = \frac{2p_{(0)}V_{(0)}}{k}\left[1 - \frac{2(1-k)}{2-k}\right] \quad \text{......} [4.2.7]$$

Making use of Equations [4.2.5] and [4.2.6]:

$$\left(\dot{m}\frac{1}{A_{(e)}}\right)^2 = \frac{x_{(e)}^{2(1-k)}}{V_{(0)}}\left[\frac{2p_{(0)}}{k}\left(1 - x_{(e)}^k\right)\right]$$

$$= \frac{x_{(e)}^{2(1-k)}}{V_{(0)}}\left[\frac{2p_{(0)}}{k}\left(1 - \frac{2(1-k)}{2-k}\right)\right]$$

$$\therefore \left(\dot{m}\frac{1}{A_{(e)}}\right)^2 = \frac{2p_{(0)}x_{(e)}^{2(1-k)}}{(2-k)V_{(0)}} \quad \text{......} [4.2.8]$$

Speed Up Factor

Equation [4.2.4] divided by Equation [4.2.7] gives the ratio $\frac{v}{v_{(e)}}$ of the velocity v at any point, where the cross section is A to the velocity $v_{(e)}$ at the muzzle, where the cross section is $A_{(e)}$. This ratio is known as the speed up factor and is given by:

$$\frac{v}{v_{(e)}} = \sqrt{\frac{(2-k)(1-x^k)}{k}} \quad \text{......} [4.2.9]$$

Divergence

The ratio of the cross sectional area A, at any point to the cross sectional area $A_{(e)}$ at the muzzle is the divergence, denoted by δ. This is obtained by dividing Equation [4.2.8] by [4.2.5].

$$\left(\dot{m}\frac{1}{A_{(e)}}\right)^2 = \frac{2p_{(0)}x_{(e)}^{2(1-k)}}{(2-k)V_{(0)}}$$

Divided by:

$$\left(\dot{m}\frac{1}{A}\right)^2 = \frac{x^{2(1-k)}}{V_{(o)}}\left[\frac{2p_{(0)}}{k}\left(1-x^k\right)\right]$$

Gives divergence:

$$\delta = \frac{A}{A_{(e)}} \quad \text{[4.2.10]}$$

For the condition that gas velocity in the chamber is negligible, divergence becomes a function solely of the pressure ratio, in other words, the relation between the speed up factor and divergence is universally acceptable regardless of other conditions.

Now the plot of divergence against speed up factor can be obtained for all values of the pressure ratio $\frac{p_{(e)}}{p_{(0)}} = x_{(e)}$ from the muzzle to the point $\frac{p_{(a)}}{p_{(0)}} = x_{(a)}$, when the gas pressure falls to atmospheric.

Example 4.2.1

Plot the divergence-speed up factor curve given the following initial conditions:

Pressure at the muzzle at projectile exit: 0.15 GPa
Volume of the chamber and barrel combined: 0.02224 m^3
Mass of charge: 3.62 kg
Ratio of specific heats for the gas: 1.26.

Solution: Example 4.2.1

The computation follows the following stages:

- Initial specific volume is calculated by dividing the initial volume by the charge mass.
- Pressure ratio at the muzzle is computed using equation [4.2.6].
- Pressure ratio is structured as an array from value at the muzzle to the ratio of atmospheric pressure to initial pressure.
- Speed up factor is computed using Equation [4.2.9].
- Divergence is computed with the help of Equation [4.2.10].
- Divergence versus speed up factor is plotted.

Computer Programme 4.2.1

```
%Computer programme to plot divergence vs speed up factor
v0=0 % chamber gas linear velocity
p0=1.514*10^8 % chamber pressure Pa
pa=1.013.*10.^5 % atmospheric pressure Pa
Vt=.02224 % chamber plus barrel volume m^3
mc=3.62% charge mass kg
V0=Vt./mc % specific volume in chamber
gamma=1.26 % ratio of specific heats
k=(gamma-1)./gamma
xe=(2.*(1-k)./(2-k)).^(1./k) % pressure ratio at muzzle
x=(xe:-.01:pa./p0) % pressure ratio array
v=sqrt(2.*p0.*V0./k.*(1-x.^k))
ve=sqrt(2.*p0.*V0./(2-k))
v_by_ve=v./ve % speed up factor
m_by_A=sqrt(x.^(2.*(1-k))./V0.^2.*(2.*p0.*V0./k.*(1-x.^k)+v0.^2))
m_by_Ae=sqrt(2.*p0.*xe.^(2.*(1-k))./((2-k).*V0))
A_by_Ae=m_by_Ae./m_by_A % divergence
plot(A_by_Ae,v_by_ve)
```

Results

Specific volume in chamber: 0.0061 m^3/kg
Pressure ratio at muzzle: 0.5531

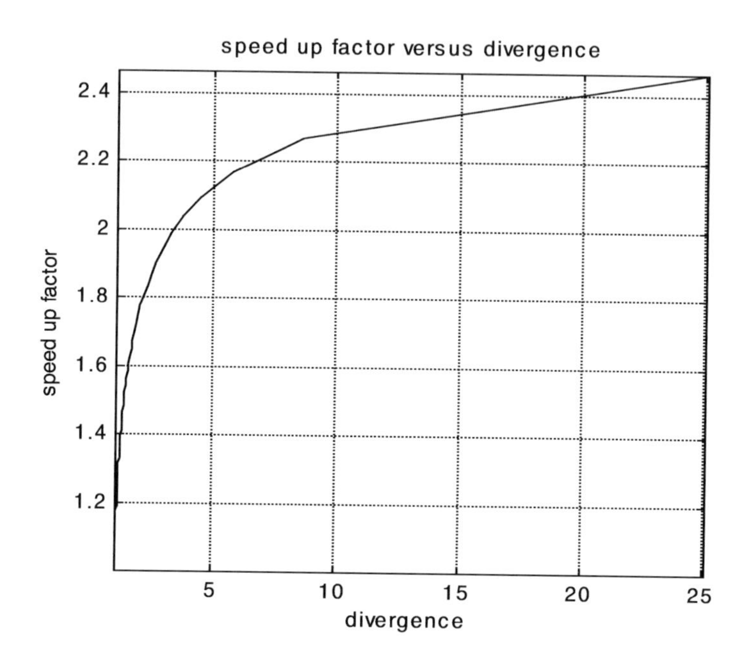

Fig 4.2.2: Example 4.2.1: Divergence versus speed up factor

4.3: Construction & Theory of Muzzle Brakes

Construction of a Muzzle Brake

A muzzle brake is a short cylindrical attachment mounted on the barrel at the muzzle end. It has a centrally bored hole through which the projectile passes, called the projectile passage. It also has one or more side openings called ports, the breech facing surfaces of the ports are called baffles. Muzzle brakes usually do not have ports on the underside because of negative effects of obscuration caused by flying debris and gas reflection off the ground. To allow symmetrical loading, of the muzzle brake, ports on the top are also avoided. Muzzle brakes may be screwed onto the barrel, with screw threading opposite in direction to that of the rifling, or may be integral with the barrel. Anti rotation arrangements with locking devices are always provided.

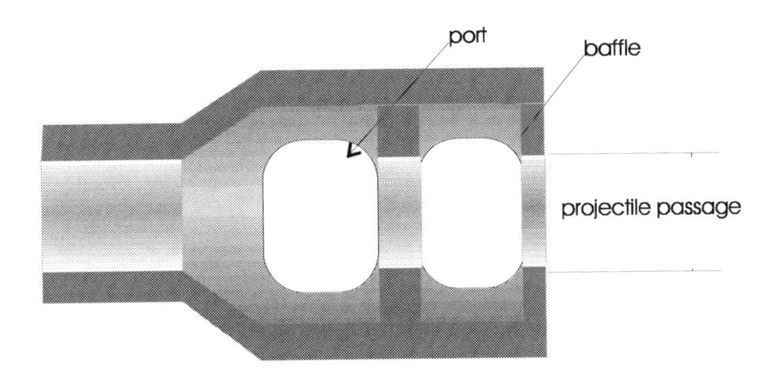

Fig 4.3.1: Cross section through a muzzle brake

Propellant gases at identical velocity to that of the projectile follow the projectile as it exits the muzzle. In the absence of any guidance, the gases expand and their pressure falls to atmospheric. When a muzzle brake is present, the projectile restricts flow of the gases through the projectile passage. The gases are now diverted to the side ports where they strike on the baffles, inducing a thrust in the direction opposite to that of recoil. This thrust times the duration for which the gases strike on the baffles generates an impulse which reduces the momentum of the recoiling parts. The muzzle brake action comes into play only at the onset of the after effect period, hence it can only reduce to some extent, the recoil energy imparted to the recoiling parts during the period under question.

Theory of Gas Deflection

The gas flow in a muzzle brake is treated as mentioned by the one dimensional nozzle theory with some simplifying assumptions as mentioned earlier. The assumption that the process is frictionless and non turbulent is found unjustified and is compensated for by the introduction of an empirical correction factor. The assumption that gas fills the brake entirely is true only for semi angles of the conical nozzle of less than 30 degrees. Smaller angles up to 15 degrees result in longer and heavier muzzle brakes which may not be acceptable. A compromise solution is possibly arrived at around 20 degrees of semi cone angle of the muzzle brake.

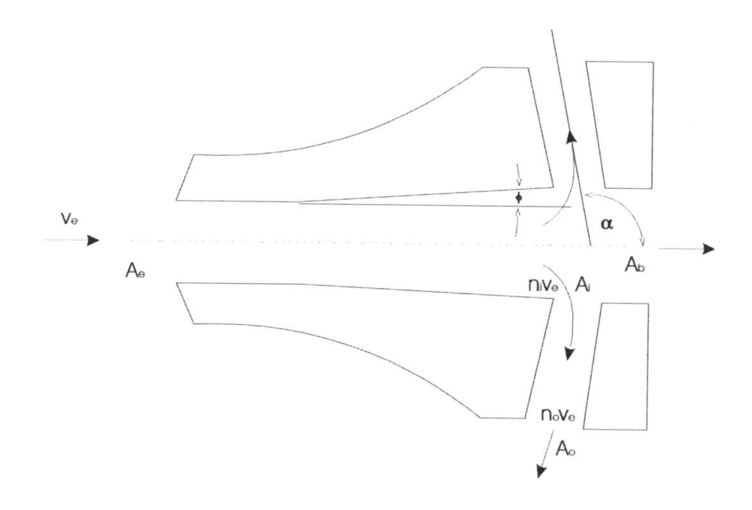

Fig 4.3.2: Gas flow through a muzzle brake

v_e: gas velocity at muzzle
n_i: speed up factor from muzzle to inlet of first port.
n_o: speed up factor from muzzle to baffle exit.
A_e: muzzle cross sectional area
A_i: port inlet cross sectional area
A_o: port outlet cross sectional area

A_b: cross sectional area of projectile passage
α: divergence angle with respect to direction of flow
φ: semi cone angle of the muzzle brake

Theory of a Single Port Muzzle Brake

Fig 4.3.2 depicts a single port muzzle brake. The basic theory of a single port muzzle brake will easily be extended to muzzle brakes with multiple ports further in the chapter.

The quantity of gas passing through the port will depend on the ratio of cross sectional area of the port to the total cross sectional area available for gas to flow. The proportion of gases flowing through the port is given by:

$$a_i = \frac{A_i}{A_i + A_b} \qquad [4.3.1]$$

The proportion of gases flowing through the projectile passage is given by:

$$a_b = \frac{A_b}{A_i + A_b} \qquad [4.3.2]$$

The divergence from the muzzle to the port inlet is given by:

$$\delta_i = \frac{A_i + A_b}{A_e} \qquad [4.3.3]$$

The divergence at the port outlet is given by:

$$\delta_o = \delta_i \frac{A_o}{A_i} \qquad [4.3.4]$$

With reference to Fig 4.2.2, it is seen that the gases would reach maximum velocity at a divergence of around 25. However the size of the brake required to achieve this divergence would be enormous. Divergences of up to 5 are viable.

Speed Up Factor

The speed up factor of a flow passage is the ratio of outlet to inlet velocity of a flow passage. Here n_i is the ratio of the gas velocity at the muzzle to the gas velocity at the inlet of the port and n_o is the ratio of the gas velocity at the port outlet to the velocity at the muzzle.

Thrust on the Baffle

If $\dot{m}$ is the mass rate of gas flow at the muzzle, then the mass rate of flow striking the baffle is given by:

$$\dot{m}_p = a_i \dot{m}$$

The thrust F_b on the baffle in unit time is equal to the change in momentum of the gas flowing through the port.

$$F_b = a_i \dot{m}(v_i - v_o Cos\alpha) \qquad [4.3.5]$$

We have:

$v_i = n_i v_e$ and $v_o = n_0 v_e$

Hence:

$$F_b = a_i \dot{m} v_e (n_i - n_o Cos\alpha) \qquad [4.3.6]$$

Or:

$$F_b = \dot{m} v_e \lambda_u \qquad [4.3.7]$$

$$\lambda_u = a_i (n_i - n_o Cos\alpha) \qquad [4.3.8]$$

Or:

$$F_b = F_m \lambda_u \qquad [4.3.9]$$

λ_u: uncorrected speed up factor.
$\dot{m}v_e = F_m$: muzzle thrust which is the muzzle pressure times the bore cross-section

Multiple Ports

The theory of gas flow through a muzzle brake with one port can be extended to muzzle brakes with multiple ports as follows:

For the first port:

A_{i1}: port inlet cross sectional area
A_{o1}: port outlet cross sectional area
A_{b1}: cross sectional area of projectile passage
α_1: gas deflection angle for the first port

For the second port:

A_{i2}: port inlet cross sectional area
A_{o2}: port outlet cross sectional area
A_{b2}: cross sectional area of projectile passage
α_2: gas deflection angle for the second port

Ratio of the quantity of gases entering the inlet of the first port:

$$a_{i1} = \frac{A_{i1}}{A_{i1} + A_{b1}}$$

Ratio of the quantity of gases entering the first projectile passage:

$$a_{b1} = \frac{A_{b1}}{A_{i1} + A_{b1}}$$

Divergence to the inlet of the first port:

$$\delta_{i1} = \frac{A_{i1} + A_{b1}}{A_e}$$

Divergence to the outlet of the first port:

$$\delta_{o1} = \delta_{i1} \frac{A_{o1}}{A_{i1}}$$

The speed up factors n_{i1} corresponding to divergence δ_{i1} and n_{o1} corresponding to δ_{o1} can now be computed from Equations [4.2.4] and [4.2.7] and Equations [4.2.8] and [4.2.5] or read off Fig 4.2.2. The uncorrected speed up factor for the first port becomes:

$$\lambda_{u1} = a_{i1}\left(n_{i1} - n_{o1} Cos\alpha_1\right)$$

Ratio of the quantity of gases entering the inlet of the second port:

$$a_{i2} = a_{b1} \frac{A_{i2}}{A_{i2} + A_{b2}}$$

Ratio of the quantity of gases entering the second projectile passage:

$$a_{b2} = a_{b1} \frac{A_{b2}}{A_{i2} + A_{b2}}$$

Divergence to the inlet of the second port:

$$\delta_{i2} = \delta_{i1} \frac{A_{i2} + A_{b2}}{A_{b1}}$$

Divergence to the outlet of the second port:

$$\delta_{o2} = \delta_{i2} \frac{A_{o2}}{A_{i2}}$$

The speed up factors n_{i2} corresponding to divergence δ_{i2} and n_{o2} corresponding to δ_{o2} can now be computed from Equations [4.2.4] and [4.2.7] and Equations [4.2.8] and [4.2.5] or read off Fig 4.2.2. The uncorrected speed up factor for the second port becomes:

$$\lambda_{u2} = a_{i2}\left(n_{i2} - n_{o2} Cos\alpha_2\right)$$

Generally for the nth port:

Ratio of the quantity of gases entering the inlet of the nth port:

$$a_{in} = a_{bn-1} \frac{A_{in}}{A_{in} + A_{bn}} \quad \text{[4.3.10]}$$

Ratio of the quantity of gases entering the nth projectile passage:

$$a_{bn} = a_{bn-1} \frac{A_{bn}}{A_{in} + A_{bn}} \quad \text{[4.3.11]}$$

Divergence to the inlet of the nth port:

$$\delta_{in} = \delta_{in-1} \frac{A_{in} + A_{bn}}{A_{bn}} \quad \text{[4.3.12]}$$

Divergence to the outlet of the nth port:

$$\delta_{on} = \delta_{in} \frac{A_{on}}{A_{in}} \quad \text{[4.3.13]}$$

The speed up factors n_{in} corresponding to divergence δ_{in} and n_{on} corresponding to δ_{on} can now be computed from Equations [4.2.4] and [4.2.7] and Equations [4.2.8] and [4.2.5] or read off Fig 4.2.2. The uncorrected speed up factor for the nth port becomes:

$$\lambda_{un} = a_{in}\left(n_{in} - n_{on} Cos\alpha_n\right) \quad [4.3.14]$$

Where a_n is the gas deflection angle of the nth port.

Correction for Friction & Turbulence: C_λ

An empirical correction factor of $C_\lambda = 1.5$ is applied to the uncorrected speed up factor to compensate for friction and turbulence effects which were earlier ignored. The speed up factor corrected for friction and turbulence is:

$$\lambda_n = \frac{\lambda_u}{C_\lambda} \quad [4.3.15]$$

Overall Speed-up Factor

At this stage, the overall speed up factor for the nth port is given by:

$$\lambda = \sum_1^n \lambda_n \quad [4.3.16]$$

Corrections to Thrust

Correction to Thrust for Obstruction by Projectile Passing through Projectile Passage: C_n

A correction is applied to the muzzle thrust to cater for the increased flow through the port, due to the closing of the projectile passage as the projectile passes through it.

This is given by:

$$C_n = \left[\sum \lambda_{n-1} + \frac{\lambda_n (A_{in} + A_{bn})}{A_{i_n}}\right]\lambda_n^{-1} \quad \text{[4.3.17]}$$

Correction Factor for Inaccuracies of Thrust Computation: C_t

When the thrust at the muzzle is computed, using the Modified Huginot Theory, a correction factor of 1.7 is applied to compensate for low computed values of thrust at the beginning of the after effect period.

Finally:

$$C_t = 1.7 C_n \quad \text{[4.3.18]}$$

The expression for total thrust is obtained by correcting the muzzle thrust. Equation [4.3.9] is modified to:

$$F_b = F_m C_t \lambda \quad \text{[4.3.19]}$$

Conversion of Chamber Pressure to Muzzle Pressure

The chamber pressure at the time of shot exit is obtained from the internal ballistics solution.

This pressure is converted to the muzzle pressure using the relation:

$$p_{mz} = p_{(e)}\left(1 - \frac{m_c}{2m_p + m_c}\right) \quad \text{[4.3.20]}$$

p_{mz}: muzzle pressure at the instant of projectile exit
$p_{(e)}$: chamber pressure at the instant of projectile exit
m_c: charge mass
m_p: projectile mass

Thrust on a Baffle

The maximum instantaneous thrust on the muzzle brake occurs when the projectile just clears the first baffle.

Assuming that the thrust is distributed amongst the baffles in direct proportion to the ratio of each individual overall speed up factor to the overall speed up factor, thrust on the nth baffle is given by:

$$F_n = F_b \frac{\lambda_n}{\lambda} \quad [4.3.21]$$

Example 4.3.1

Compute the overall individual speed up factors for a 130 mm gun with a 3 baffle muzzle brake given the following data:

Muzzle velocity: 930 m/s
Angle of deflection of gas: 120°
Friction & turbulence factor: 1.5
Projectile passage area: 0.015 m^2
Port inlet area (for all ports): 0.007 m^2
Port exit area (for all ports): 0.009 m^2
Mass of charge: 3.925 Kg
Mass of projectile: 14.08 Kg
Chamber pressure at projectile exit: 37.0 M Pa

Solution: Example 4.3.1

This is conveniently solved by constructing an Excel spreadsheet as follows:

- Ratio of port inlets to total area is calculated using Equation [4.3.10]
- Ratio of projectile passages to total area is calculated using Equation [4.3.11]
- Divergence to port inlets is calculated using Equation [4.3.12]

- Ratio port exit to inlet area is calculated by dividing the respective port outlet cross sectional area by the port inlet cross sectional area.
- Divergence at port exit is calculated from equation [4.3.13]
- Speed up factor corresponding to $\delta i(n)$ is obtained from Fig 4.2.2
- Speed up factor corresponding to $\delta e(n)$ is obtained from Fig 4.2.2
- Uncorrected speed up factor is calculated using Equation [4.3.14]
- Corrected overall speed up factor is obtained with the help of Equation [4.3.16]
- Correction to thrust is obtained from Equation [4.3.18].
- Chamber pressure at projectile exit is converted to muzzle pressure with the help of Equation [4.3.20].
- Net thrust is obtained from Equation [4.3.19].
- Thrust on individual baffles is computed using Equation [4.3.21].

Worksheet for Speed Up Factor & Thrust Computation: Formulae

	A	B	C	D
1	**Worksheet to compute speed up factors & thrust on baffles: 3 port muzzle brake**			
2	D	0.13	α	120
3	Cλ	1.5	Cosα	=COS($D$2/180*PI())
4	Ae	=B2^2/4*PI()	pe	=37*10^6
5	mc	3.925	pmz	=D4*(1-B5/(2*B6+B5))
6	mp	14.08		
7	Ab	0.015	0.015	0.015
8	Ao	0.014	0.014	0.014
9	Ai	0.007	0.007	0.007
10	ai	=B9/(B7+B9)	=C9/(C7+C9)	=D9/(D7+D9)
11	ab	=B7/(B7+B9)	=C7/(C7+C9)	=D7/(D7+D9)
12	Ae/Ai	=B8/B9	=C8/C9	=D8/D9
13	δi	=B9/(B7+B9)	=C9/(C7+C9)*B11	=D9/(D7+D9)*C11*C11
14	δb	=B7/(B7+B9)	=C7/(C7+C9)*C11	=D7/(D7+D9)*D11*D11
15	δ	=(B9+B7)/$B$4	=(C9+C7)/B7*B15	=(D9+D7)/C7*C15
16	δe	=B15*B12	=C15*C12	=D15*D12
17	ni	1.94	2.25	2.43
18	no	2.05	2.32	2.47
19	noCosα	=B18*$D$3	=C18*$D$3	=D18*$D$3
20	λn	=B13*(B17-B19)/$B$3	=C13*(C17-C19)/$B$3	=D13*(D17-D19)/$B$3
21	λ	=B20	=B21+C20	=C21+D20
22	Cn	=B20/B10/B21	=(B21+C20/C10)/C21	=(C21+D20/D10)/D21
23	Ct	=B22*1.7	=C22*1.7	=D22*1.7
24	Fb	=$D$5*$B$4*B21*B23	=$D$5*$B$4*C21*C23	=$D$5*$B$4*D21*D23
25	Fn	=B24*B20/B21	=C24*C20/C21	=D24*D20/D21

Fig 4.3.3: Worksheet for computation of speed up factors & thrust on baffles of Example 4.3.1.

Worksheet for Speed Up Factor & Thrust Computation: Results

	A	B	C	D
1	**Worksheet to compute speed up factors & thrust on baffles: 3 port muzzle brake**			
2	D	0.130	α	120.000
3	Cλ	1.500	Cosα	-0.500
4	Ae	0.013	pe	3.70E+07
5	mc	3.925	pmz	3.25E+07
6	mp	14.08		
7	Ab	0.015	0.015	0.015
8	Ao	0.014	0.014	0.014
9	Ai	0.007	0.007	0.007
10	ai	0.318	0.318	0.318
11	ab	0.682	0.682	0.682
12	Ae/Ai	2.000	2.000	2.000
13	δi	0.318	0.217	0.148
14	δb	0.682	0.465	0.317
15	δ	1.657	2.431	3.565
16	δe	3.315	4.862	7.131
17	ni	1.940	2.250	2.430
18	no	2.050	2.320	2.470
19	noCosα	-1.025	-1.160	-1.235
20	λn	0.629	0.493	0.361
21	λ	0.629	1.122	1.484
22	Cn	3.143	1.942	1.522
23	Ct	5.343	3.301	2.587
24	Fb	1.45E+06	1.60E+06	1.65E+06
25	Fn	1.45E+06	7.02E+05	4.03E+05

Fig 4.3.4: Worksheet to compute Speed Up Factor & thrust on baffles of muzzle brake of Example 4.3.1.

5

Supporting Structures

5.1: The Superstructure

Supporting Structures

The supporting structure of a heavy artillery weapon may be permanent in location or mobile. Static supporting structures are associated with coastal or air defence artillery, permanently positioned, to fulfill a limited tactical role. Mobile supporting structures are usually referred to as carriages or mountings. Generally, a carriage is associated with the supporting structure of a gun which fires off its wheels and mounting relate to supporting structures which have rigid members in contact with the ground, when prepared for firing. Much debate is possible as to the appropriateness of the terminology. Regardless of the expression used to designate the supporting structure, all supporting structures have common functions.

Functions of the Supporting Structure

The functions of the supporting structure are:

- Provision of a stable support to the gun when firing and during travel. When a gun fires it is necessary that the energy of recoil is absorbed and dissipated, in such a manner that the stability of the gun is preserved. A gun is deemed stable if the main points of contact of the supporting structure with the ground, which are usually the wheels and spades of the trails, or the tracks in the case of AFVs, remain in contact with the ground, during recoil and counter recoil. Stability is difficult to achieve. Practically, weight is a major limitation for all but static and armoured vehicle mounted guns.
- To house the components and mechanisms to enable the gun to be laid for azimuth and elevation in any desired attitude.
- To absorb the energy of recoil smoothly transmitted to it and to dissipate this energy without affecting the stability and without shift in the location of the gun.
- In the case of heavy weapons, to mount the automatic means of loading and ramming.
- Provide a means of attachment of the equipment with its prime mover this to include arrangements for braking and tail lighting. When coupled to the prime mover, the turning radius, when negotiating bends on the route, should conform to the realistic requirements of road geometry.
- In most modern equipments, to house the power source and transmission for limited tactical self-propulsion and to meet the power requirements of the gyroscopes and electrical components of the fire control system.

Components of an Artillery Supporting Structure

The supporting structure for an artillery gun may be referred to as a carriage. The carriage in turn consists of what is termed the superstructure and the basic structure. The super structure includes those parts, other than the gun itself which move over the bottom carriage in traverse. The basic structure provides the level foundation on which the super structure is traversed.

Tanks and Self-Propelled Artillery

Although tanks and self-propelled artillery are based for mobility on identical chassis, the similarity ends here. Tank guns are high velocity weapons intended primarily to destroy, by direct fire, adversaries' tanks of similar weight classification, at ranges in excess of that at which they themselves can be engaged. This to include all battlefield circumstances such as poor visibility and targets that are on the move. The enhanced mobility of self-propelled artillery does not alter its role. Hence self-propelled artillery is not designed to fire on the move or engage targets by direct fire in conditions of poor visibility.

From the armament aspect, the salient differences between a self propelled artillery piece and a battle tank are as follows:

- The primary ammunition of self-propelled artillery remains the HE shell. Secondary anti tank ammunition of the chemical energy type is usually included in the inventory for self-protection. The gun is therefore designed for lower muzzle velocities.
- The fire control system for direct firing is basic in character. Range finders and various sensors are not incorporated on individual weapons. Night vision capability for the gunner is not essential. The indirect fire control system is elaborate.
- Self propelled artillery guns are not designed to fire on the move and are not stabilized.
- The elevation and traversing gearing is not designed for high angular velocities or high accelerations for target tracking. Impulse loading of the gear trains as a result of vehicular turbulence is also absent.
- Since high velocities are absent, gears trains are usually manually powered and simple stops are incorporated instead of elevation and traversing buffers.
- The gun is usually maintained within a fixed arc of traverse relative to the hull to avoid firing loads on the suspension of the

vehicle. The suspension may be locked out during firing and retractable spades could be made use of.

The Super Structure

The super structure consists of the cradle and the saddle also sometimes called the top carriage.

The Cradle

Functions of the Cradle

This component of the carriage permits movement of the gun in the vertical plane, while allowing the gun rearward movement during recoil. It also serves at the static component during retardation of the gun during recoil. Its functions in better detail are as follows.

- The cradle carries the gun and all the components, which move in elevation. By virtue of the trunnions being integral to the cradle, movement of the gun in elevation is possible, yet allowing movement of the gun in the axial direction, during recoil and counter recoil, by means of slides on which guides attached to the tube slide.

- The cradle anchors the fixed components of the recoil mechanism. The recoil and recuperator piston rods or the recoil brake and recuperator cylinders are anchored to the cradle, as the case may be.

- It prevents the tube from rotating when induced to do so by the reaction torque experienced due to the rotational motion of the projectile driving band. This is achieved by grooves on the slides mating with the edges of the guides or some such similar arrangement.

- It transmits all the loads including those induced by recoil braking, jump and rifling torque to the top carriage through the trunnions.

- By virtue of the trunnions the cradle provides an accurate measure of the elevation angle of the barrel. Ideally, though not always practicable, measurement of the elevation angle is preferable at the trunnions itself. Most often for reasons of accessibility to the gunner, the sighting equipment has to be displaced from the trunnions employing parallel link mechanisms.

- The elevating pinion shaft rotates within bearing in a shaft hole in the cradle. The elevating gear box is also normally attached to the cradle.

- A recoil slide indicator also provides a measure of the recoil length, which is an indicator of the gas pressure in the recoil system recuperator.

- One end of the balancing gear is anchored to the cradle.

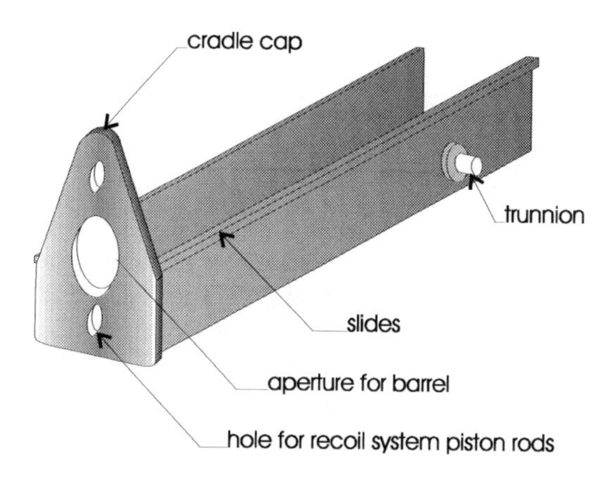

Fig 5.1.1: Trough type cradle

Components of the Cradle

Fixed Components of the Recoil System

The components of the recoil system, which are anchored to the cradle, may be considered part of the cradle itself. These may be the piston rods' in the case of fixed piston moving cylinder recoil systems or the cylinders in the case of the fixed cylinder moving piston recoil systems. In both cases placement of the components with respect to the cradle must minimize the bending moments which the cradle will be subject to.

Trunnions

The trunnions are the pivots about which the elevating parts rock. The trunnions are integral to the cradle and their disposition largely influences the moments which occur during firing. Ideally the trunnion axis must lie on the plane through the axis of the bore. This plane must also contain the centre of gravity of the recoil system. To achieve this, the recoil system is usually distributed evenly on both the upper and the lower sides of the bore axis.

Balancing Gear Pivot

One extremity of the balancing gear is attached to the cradle. The distance of this point of attachment must be as large as possible to reduce loads on the balancing gear.

Elevating Arc

The elevating arcs are rigidly attached to the cradle or the top carriage with their pitch radii at the trunnion axis. Elevating arcs are employed in pairs on either side of the cradle or top carriage in order to have symmetric loading of all components involved. In case the arcs are attached to the top carriage, the pinion shaft will perforce be housed in the cradle and vice versa. The largest pitch radius within the overall dimensional constraints of the weapon is desirable from the point of view of low elevating effort and gear tooth loads. A large arc radius also betters distribution of loads at the points of attachment to the cradle.

Types of Cradles

Ring type Cradle

This type of cradle finds favour in turreted gun assemblies. The cradle is of cylindrical shape with a bearing surface inside on which the gun slides during recoil. External brackets are provided for anchoring the recoil system components. A key is incorporated between the cylindrical portion of the barrel, which serves as the slide and the inner cylindrical surface of the cradle, the guide to prevent rotation of the barrel during firing. The advantages of the ring type cradle are:

- Bending moments on the cradle are evenly distributed.
- The requirement of slides attached to the barrel is eliminated. Hence the preparation of cylindrical surfaces specifically for this purpose is rendered unnecessary.
- It is feasible to position the centre of gravity of the cradle on the plane containing the axis of the bore.
- Alignment of the bore axis with the axis of the cradle is simpler.

The ring type cradle is not without disadvantages:

- The sliding surface of the tube has to be cylindrical and may not conform to the desired dimensions of the gun from design considerations, resulting in additional weight.
- The effect of thermal expansion of the barrel may cause seizing of the barrel in the cradle bearings. Increased clearance between the barrel and the bearing surface, to cater for this expansion, may not be acceptable when the gun is cold.

Trough Type Cradle

As the name implies, this type of cradle is similar in configuration to a drinking trough, it is made from a one-piece casting, or welded construction, strong enough to withstand the stresses, which occur between the anchoring points of the recoil assembly and the trunnions. Flat slides are provided for the movement of the gun guides. The main advantages of the trough type cradle are as follows:

- The contour of the gun barrel is affected only in that cylindrical portions are necessary for the attachment of the slides.

- Expansion of the gun barrel does not cause any problem as regards recoil movement.

However this type of cradle offers some disadvantages:

- The ideal loading condition to reduce bending moments is difficult to attain.

- Accuracy of alignment of the slides and guides during manufacture is difficult.

- Misalignment of the slides and guides during service is frequent and may lead to more complicated problems.

Concentric Type Cradle

This is similar to the ring type cradle but in this case the cradle itself houses the recoil system symmetrically distributed about the axis of the bore. Such a cradle finds favour in turreted weapon systems. With this type of cradle placement of the centre of gravity of the recoil system on the bore axis is easy, thereby removing undesirable moments at the trunnions.

Saddle or Top Carriage

The top carriage is the component which rotates on the basic structure thereby imparting traverse in the horizontal plane to the superstructure.

It has the following important functions:

- The top carriage or saddle is the primary supporting structure of the weapon. It supports the cradle through the trunnion bearings.
- It transmits all firing loads from the cradle to the bottom carriage and other supporting members.
- The top carriage serves to anchor one end of the balancing gear mechanism.
- The top carriage houses the elevating and traversing arcs in case the elevating and traversing pinions are attached to the cradle or vice versa. Trunnion and traversing bearings are provided to reduce frictional forces during traverse and elevation. The bearing being strong enough to withstand static and firing loads, while permitting free rotation of the elevating and traversing units.

The saddle consists of side arms, which support the trunnions, held together by a base plate like an inverted saddle or letter U. The bottom surface of the base plate contacts the bearing surface of the bottom carriage. The sub components of the saddle are:

- The vertical pivot pin, the vertical axis of which is the traversing axis of the superstructure of the gun.
- Horizontal traverse bearing surfaces for low friction traversing. Flat bearing surfaces of bronze fixed to the bottom carriage are generally used when the equipment does not fire on the traverse and bearing pressures are limited. In case of weapons that fire while traversing, roller bearings are preferred.
- Trunnion bearings reduce the friction force loads on the elevating mechanism. They are also the conduits for transfer of loads from the cradle to the top carriage. Trunnion bearings may be either of the shell or roller type. Roller bearings offer the advantages of low

friction coupled with self-aligning properties. However they are comparatively larger in size and costlier than shell bearings. Shell bearings are compact and cheap but do not have desirable low friction attributes.

- Removable trunnion cap squares. These are semicircular retaining components fitted over the top half of the trunnion bearings and allow easy access for lubrication of the trunnion bearings and disassembly of the cradle when required.
- Elevation and traverse stops. These are intended to limit the traverse and elevation of the top carriage and cradle to designed limits. In manually operated weapons, these consist simply of lugs on the fixed member interacting with teeth on the moving member. In powered systems, where angular accelerations are of a high order, some form of brakes or buffers are necessary to cushion impact between the fixed and movable members.
- Attachments for balancing gears and sights. The top carriage usually houses the balancing gear cylinders and shaft for mountings the sighting system. A linkage is provided between the trunnions and this shaft to correspond the rotational movement of the two.

Types of Top Carriages

Basically two types of top carriage exist. The top carriage offering unlimited traverse of 360° as in tank applications and the limited traverse associated with artillery guns. With artillery guns this traverse is usually around 30° to either side of the centre line of fire. In these case s, the deficiency in traverse is made up, somewhat, by traversing the weapon as a whole into the desired direction of fire at the time of deployment.

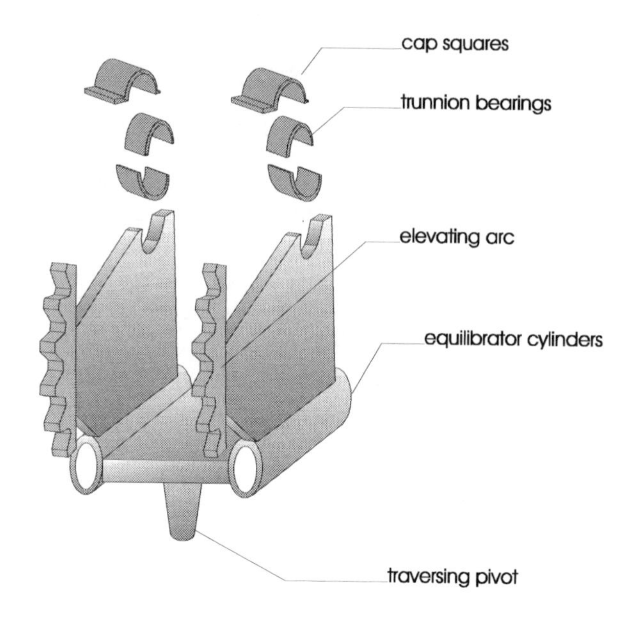

Fig 5.1.2: Components of a top carriage

5.2: The Basic Structure

Functions of the Basic Structure

The basic structure, sometimes called the bottom carriage, consists of the components which provide a rigid foundation for the weapon during firing, a base on which the superstructure can be traversed in azimuth and the vehicular part of the equipment during movement. The basic structure should ideally be light in weight, strong and symmetric about the central axis for even load sharing. Simplicity, ease of manufacture and maintenance and speed of deployment are other requirements. The functions of the basic structure are listed below:

- The basic structure provides support to the top carriage /saddle.

- It provides a recess and bearing surfaces for the pivot of the super structure for traverse of the gun.
- The basic structure and its components form the structural foundation of the weapon and transmit all loads to the ground.
- During transit the bottom carriage becomes the chassis for the equipment as a whole. It therefore incorporates vehicular components such as tracks and road wheels, pneumatic or solid wheels, suspension, braking systems both for dynamic and for parking conditions and tail and brake lighting.

The bottom carriage generally has the following components and sub assemblies:

Front Support

This is a beam like component which provides the recess for the pivot of the top carriage. It is fitted with a bearing surface at the surface of contact with the top carriage. The inner surface of the recess in the front support carries the horizontal or radial loads. In the case of split trails it houses the horizontal pin about which the axle articulates. The vertical loads are transmitted to the ground via the horizontal pin the axle tree and the wheels. The trails, hinged to the front support, are closed during transportation and spread outwards during emplacement for firing.

Axles

Axles may be either over slung or under slung. Under slung axles afford the advantages of allowing a lower silhouette, better stability but limit the elevation possible.

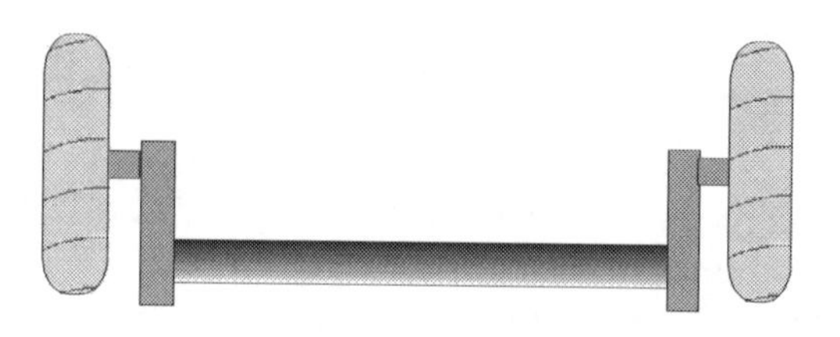

Fig 5.2.1: Under slung axle

Fig 5.2.2: Over slung axle

Suspension

A suspension system is provided for smooth travel during road movement. This consists essentially of springs and shock absorbers. The springs may be mechanical springs or torsion bars. Pneumatic springs are also in evidence. Shock absorbers dampen the oscillations of the springs. A means of locking out the suspension during firing, in order to render the structure rigid, except during traveling, is also built-in. Some self-propelled artillery may be restricted to firing only in the broadside position with respect to the hull, to avoid loading the suspension system with the firing impulse.

Brakes

Brakes are provided on the carriage to prevent movement on firing, for parking the gun on gradients, as also to provide additional braking effort over that of the prime mover during towing and to control motion of the gun when independent of the prime mover. Brakes may be internal expanding shoe type and may be hydraulic or pneumatically assisted in operation. Hand operated brakes are used during man handling, parking and during firing. Overrun brakes are controlled through hydraulic or air lines from the prime mover.

Trails

The trails are long beam like members of hollow circular or rectangular cross section, attached by hinges to the basic structure. During firing they are splayed to provide stability to the structure as seen in Chapter 3. During transportation, the trails are closed and locked together forming a rigid towing bar which hooks onto the rear of the towing vehicle. Some designs incorporate collapsible trails to reduce the turning radius. The functions of the trails are as follows:

- Transmission of firing stresses to the ground.
- Provision of a stable platform for the gun during firing, and recoil. The length of the trail has a direct bearing on the stability of the gun during recoil as was seen in Chapter 3.
- Maintaining the weapon steady in the firing position.
- To connect the gun to the tractor for towing

Types of trails

- Pole Trail. The pole trail is normally associated with vintage cannon. It consisted of a single member, which had a towing eye at one end and attached directly to the tower. It provided stability over a limited arc of fire or small top traverse, if any. The barrel was elevated by a screw type mechanism between the trail and the breech end of the gun. Elevation of the barrel was however restricted due to presence of the trail under the rear end or breech mechanism of the gun.

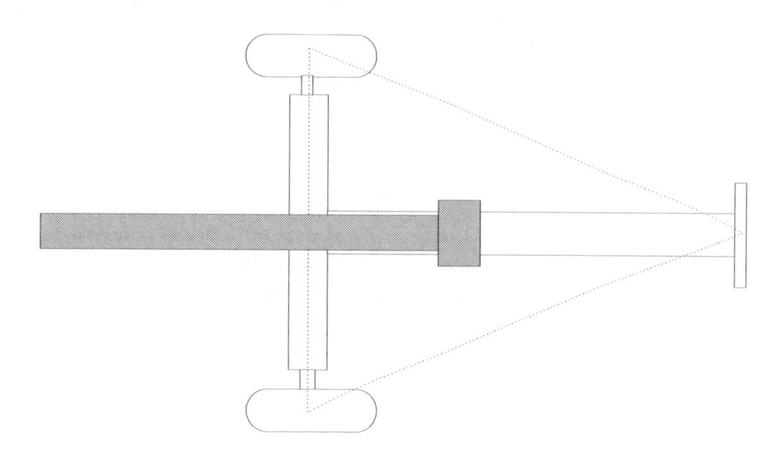

Fig 5.2.3: Pole trail showing foundation figure

- Box Trail. This is a modified single or pole trail. This consists of two stability members joined at one end. It may be parallel sided, splayed or bowed. It affords a poor foundation figure, hence top traverse is limited to about 150-200 mils, due to fouling with the side members. However if used in conjunction with a platform and tie bars, an all round traverse is possible. Elevation to angles $> 45°$ is possible. The stresses being shared between the two members, it does not offer much in terms of weight reduction as compared to the pole trail.

- Split Trails. This consists of two independent members which may be straight or hinged at the mid point of each trail, to reduce the overall length of the equipment. The overall length of the equipment is limited by the turning radius while negotiating turns in conjunction with the tractor. The trails open out for firing affording a wide foundation figure. Better top traverse is possible, but all round traverse is possible only by physically rotating the equipment on its wheels. This may call for re-alignment, or recording, of the gun for azimuth. High angle fire is feasible. The trails are individually heavy to withstand stresses which, depending on the line of action of the firing loads due to traverse, may not be equally shared between them. Some form of articulation is a mandatory for bottom carriages employing split trails.

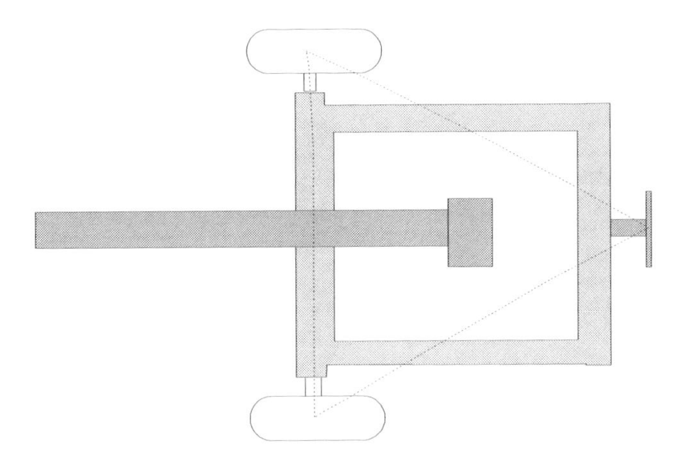

Fig 5.2.4: Box trail showing foundation figure

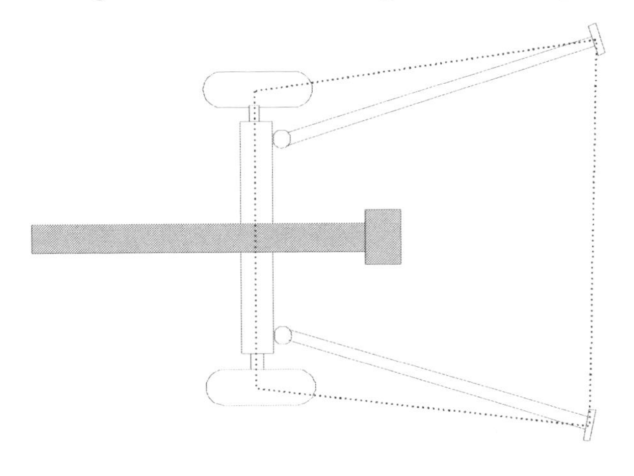

Fig 5.2.5: Split trails showing foundation figure

Spades

The function of the spades is to prevent rearward movement of the gun when firing. In the case spades and of platforms, shoes or surfaces on the spades are provided to prevent digging into the ground, of the spades or platform.

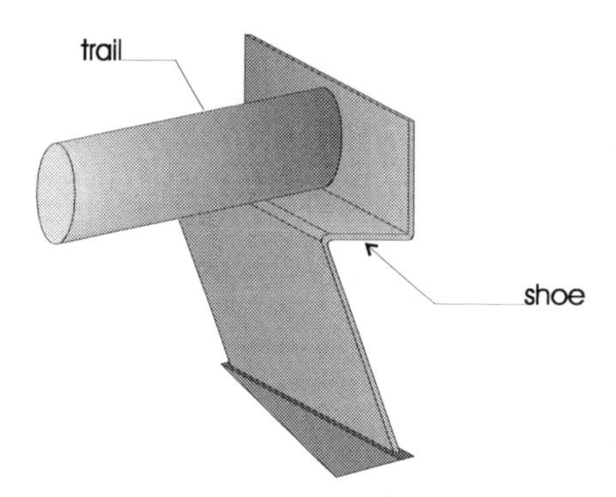

Fig 5.2.6: Spade

Articulation

This is a mechanical arrangement incorporated in bottom carriages of equipments with split trails. This arrangement ensures consistent four point contact, with the ground and hence a stable platform. Four-point contact is maintained on uneven ground only by arranging relative movement in transverse planes between the axles and the trails. The movement provided by articulation is limited to a few degrees, but is sufficient to compensate for small unevenness in the ground. The articulation is locked out on closing the trails to prevent damage during travel.

Methods of Articulation

Numerous methods of articulation have been devised and are in use; some common types of articulation are listed_below:

- Longitudinal Pivot. Here the pin passes through the axle and saddle support. Each trail is joined to the saddle support by a vertical pin. The saddle support pivots about the longitudinal pin, while the wheels are in ground contact. When the trails are closed, the projection in the trails bear against the axle thereby locking the trails for travel.

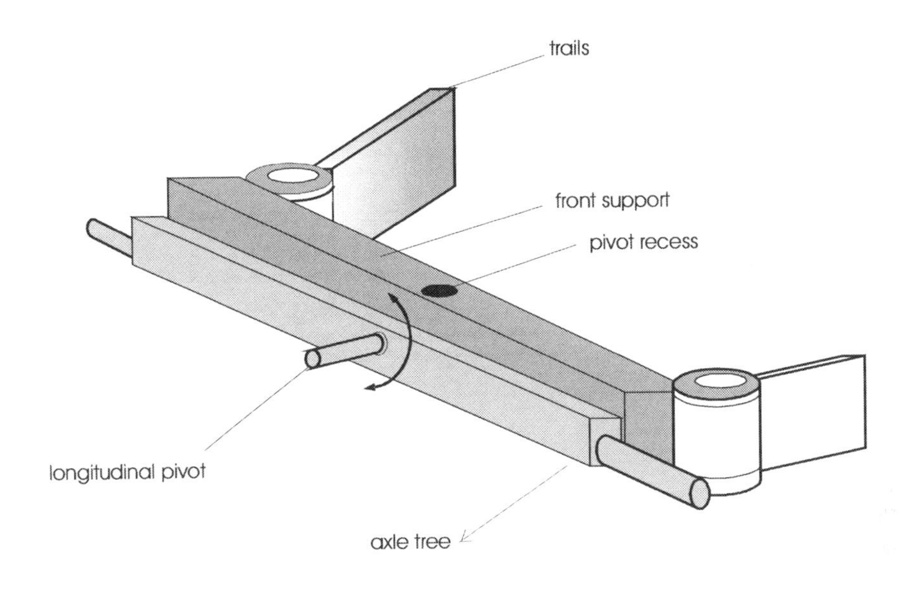

Fig 5.2.7: Longitudinal pivot articulation

- Rocking Arm. The rocking arms pivot about each end of the saddle support. The trails are attached to the rocker arms by vertical pins. The tie rod pivoted in the middle of the saddle support transmits articulation as also limits it.

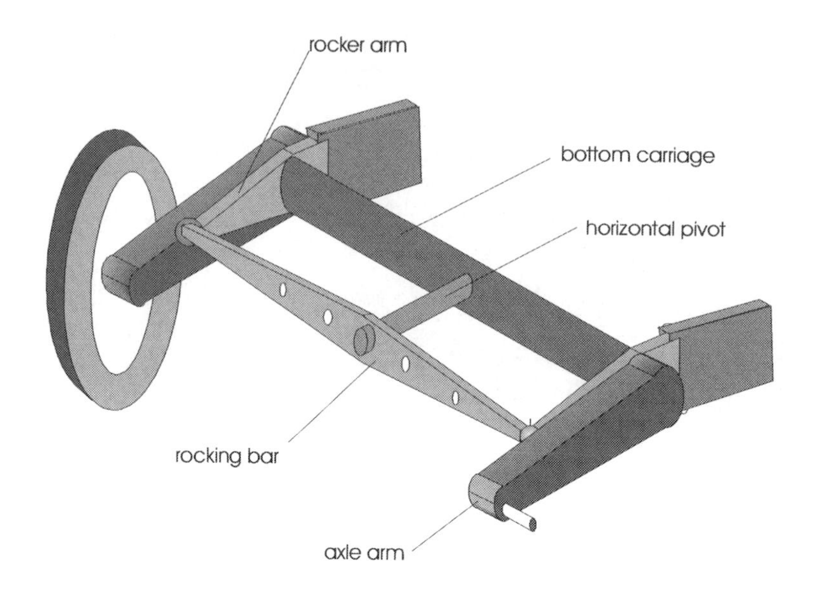

Fig 5.2.8: Rocking arm articulation

- Miscellaneous Methods. Other methods of articulation include:

    - Ball and socket.
    - Parallelogram joint.
    - Bevel gearbox.
    - Horizontal spring.
    - Bevel segment.
    - Connecting link.

5.3: Loads on the Structure

Loads on the Cradle

By virtue of its function, the cradle is subject to loads and torques which it in turn transmits to the top carriage. These include the recoil braking force, the balancing gear force, the reaction force at the pinion or arc of the elevating gear train and the rifling torque.

Recoil Force

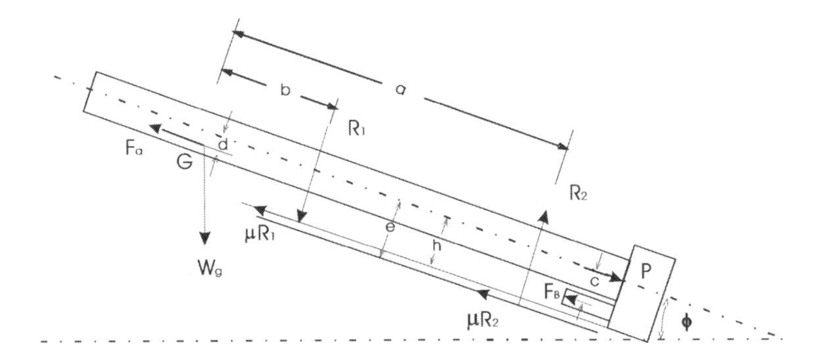

Fig 5.3.1: Forces on recoiling parts

With reference to Fig 5.3.1, the accelerating force of the recoiling parts is given by:

$$F_a = P + W_g Sin\phi - F_{Bnet} \quad [5.3.1]$$

P: propellant gas force
Wg: weight of recoiling parts
ϕ : angle of elevation
F_{Bnet}: net braking force, obtained as detailed in Chapter 3

Also:

The net braking force is the sum of the recoil system braking force plus the friction force at the slides given by:

$F_{Bnet} = F_B + F_F$; F_B: recoil system braking force, F_F: friction force at the slides

$$F_F = \mu(R_1 + R_2) \quad [5.3.2]$$

Taking moments about the point of intersection of the line of action of R_2 and the surface of the slides:

$$(e-c)F_B + eP - (e-d)(F_a - W_g Sin\varphi) - aW_g Cos\varphi - (e-h)\mu R_1 - (a-b)R_1 = 0 \quad [5.3.3]$$

The values of R_1, R_2, which are the reactions to the moment of the recoiling forces, having been established from Equations [5.3.2] & [5.3.3], it is now possible to compute the maximum bearing pressure on the slides. In the case of continuous contact between the guides and the slides, these reactions can be assumed to have a triangular load distribution as depicted in Fig 5.3.2.

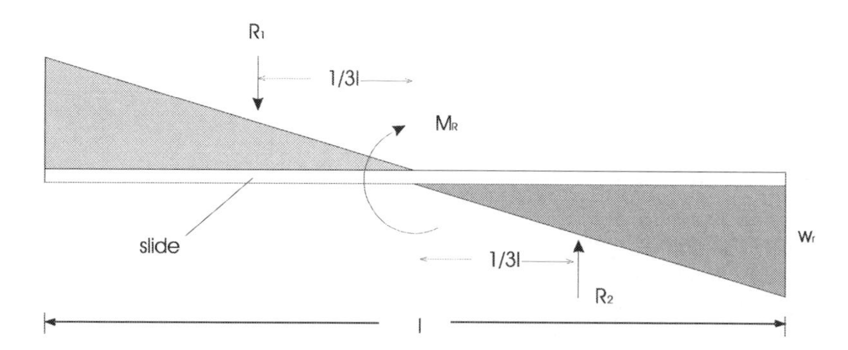

Fig 5.3.2: Load distribution on continuous slides of trough type cradle

Maximum Load due to Recoil Forces & Weight Component of Recoiling Parts

Taking the triangular area of Fig 5.3.2, which represents the load distribution due to the larger reaction, either R_1 or R_2, say R_2, the maximum load per unit length w_r, due to recoil forces, on a single slide is given by:

$$\frac{R_2}{2} = \frac{1}{2}\left(\frac{l}{2}\right)w_r$$

Or:

$$w_r = \frac{2R_2}{l} \qquad [5.3.4]$$

Reaction due to Rifling Torque

One of the functions of the cradle is to transmit the rifling torque. The basic torque equation is given by:

$$T = Ia \qquad [5.3.5]$$

T: rifling torque
I: moment of inertia of the projectile
a : angular acceleration of the projectile

We have:

$$P = m_p a \text{ and } Tan\theta = \frac{\pi}{N}$$

P: propellant gas force
m_p: projectile mass
a: linear acceleration of the projectile

θ : angle of rifling
N: twist of rifling

It follows that the peripheral acceleration of the projectile is:

$$a_p = aTan\theta \quad \text{[5.3.6]}$$

Or:

$$a_p = \frac{p\pi}{m_p}\left(\frac{D}{2}\right)^2 \frac{\pi}{N} = \frac{p\pi^2 D^2}{4m_p N} \quad \text{[5.3.7]}$$

The angular acceleration of the projectile is related to the peripheral acceleration by the expression:

$a = \dfrac{2a_p}{D}$, hence:

$$a = \frac{p\pi^2 D}{2m_p N} \quad \text{[5.3.8]}$$

Generally $k^2 = 0.6\left(\frac{D}{2}\right)^2$, k being the radius of gyration of the projectile.

Hence:

$I_p = 0.6m_p\left(\frac{D}{2}\right)^2$, substituting from Equations [5.3.5] to [5.3.8]:

$$T = \frac{0.6\pi^2\left(\frac{D}{2}\right)^3 p}{N}$$

With reference to Fig 5.3.3, the load on each trunnion due to the rifling torque is:

$$F_t = \frac{T}{D_t} \quad \text{[5.3.9]}$$

The load on each slide is:

$$F_{sl} = \frac{T}{D_{sl}} \quad \text{[5.3.10]}$$

D_t, D_{sl}: moment arms as detailed in Fig 5.3.3
The torque is transmitted from the guides to the slides in the form of vertical forces with a moment arm equal to the distance from their lines of action to the bore axis, Fig 5.3.3 refers.

The load distribution, due to rifling torque, on the slides per unit length is given by:

$$w_t = \frac{F_{sl}}{l} \quad \text{[5.3.11]}$$

The maximum load distribution on the slides becomes $w = w_r + w_t$ and the maximum pressure on the slides is given by:

$$p_{sl} = \frac{w}{b} \quad \text{[5.3.12]}$$

b: width of the bearing surface of the slides.

The maximum pressure on the slides for intermittent motion should not exceed 3.5 M Pa. If in excess of this magnitude it would result in severe wear and tear of the contact surfaces.

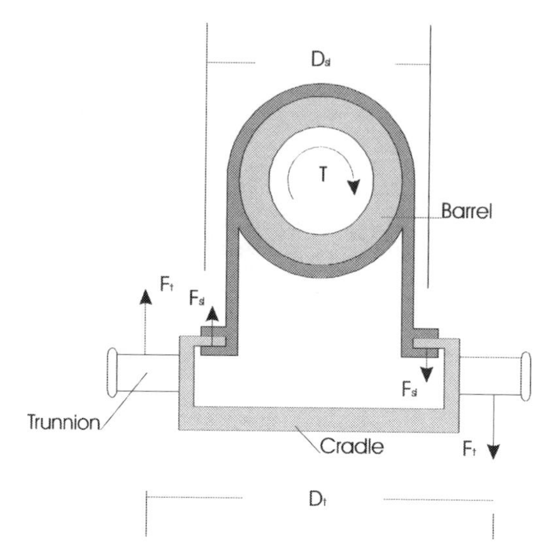

Fig 5.3.3: Load on slides due to rifling torque

Load due to Balancing Gear

With reference to Fig 5.3.4 and assuming, for simplicity, that the center of gravity of the both the recoiling parts and the cradle lie on the axis of the bore. Also that the axis of the trunnions lies on the plane containing the bore axis. The normal load on the cradle due to the balancing gear is calculated by taking moments about the trunnions.

$$P_e e Cos\lambda = W_g a Cos\phi + W_c c Cos\phi$$

From which:

$$P_e = \frac{W_g a Cos\phi + W_c c Cos\phi}{e Cos\lambda} \quad \text{[5.3.13]}$$

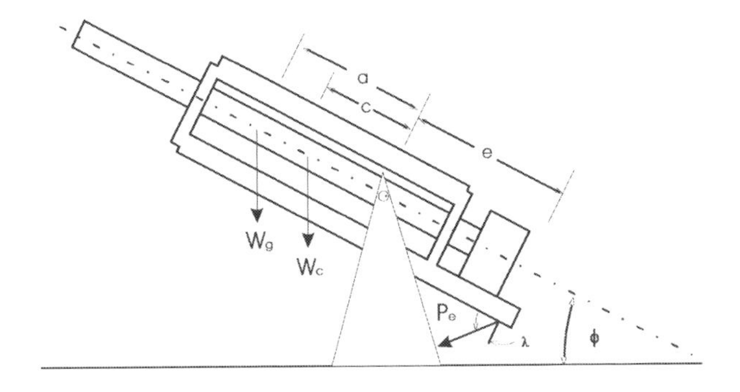

Fig 5.3.4: Load on cradle due to balancing gear

Elevating Pinion Load

With reference to Figs 5.3.4 and 5.3.5, the reaction at the elevating gear pinion R_p is found by taking moments about the trunnions:

$$R_p r_e Cos\beta - P_e e Cos\lambda - cW_c Cos\phi - W_g a Cos\phi = 0$$

r_e: elevating arc radius

From which the load on the elevating pinion R_p can be determined.

Loads on the Top Carriage

The loads on the trunnions are determined by summation of the forces in the direction of the axis of the bore and normal to it:

Summation of forces normal to the bore axis gives:

$$F_{t_N} = P_e Cos\lambda - R_p Cos\beta + W_g Cos\phi + W_c Cos\phi \quad [5.3.14]$$

$$F_{t_A} = F_a + W_g Sin\phi + W_c Sin\phi - P_e Sin\lambda - R_p Sin\beta \quad [5.3.15]$$

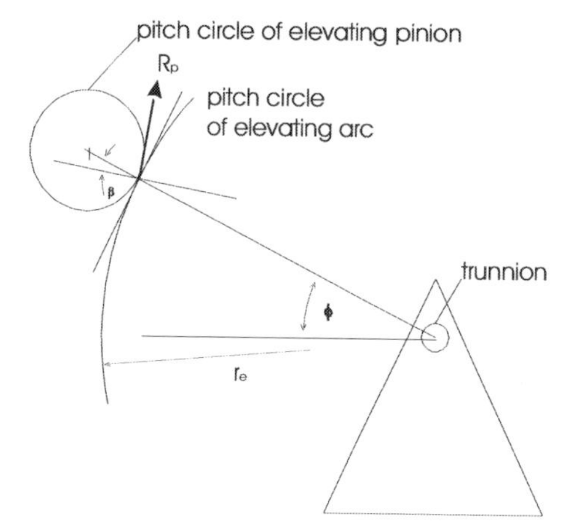

Fig 5.3.5: Load on elevating pinion

Loads on the Bottom Carriage

The forces, which are applied to the bottom carriage, are the trunnion loads, the balancing gear loads, the load on the elevating arc and the weight of the top carriage. Necessarily, the weight of the top carriage itself is taken into account. Assuming that the top surface of the bottom carriage is horizontal, the loads mentioned can be resolved into their respective vertical and horizontal components.

With reference to Fig 5.3.6, the loads on the bottom carriage are resolved into their horizontal and vertical components as follows:

The resolved trunnion load is:

$$H_t = F_{t_A} Cos\varphi + F_{t_N} Sin\varphi \quad [5.3.16]$$

$$V_t = F_{t_A} Sin\varphi - F_{t_N} Cos\varphi \quad [5.3.17]$$

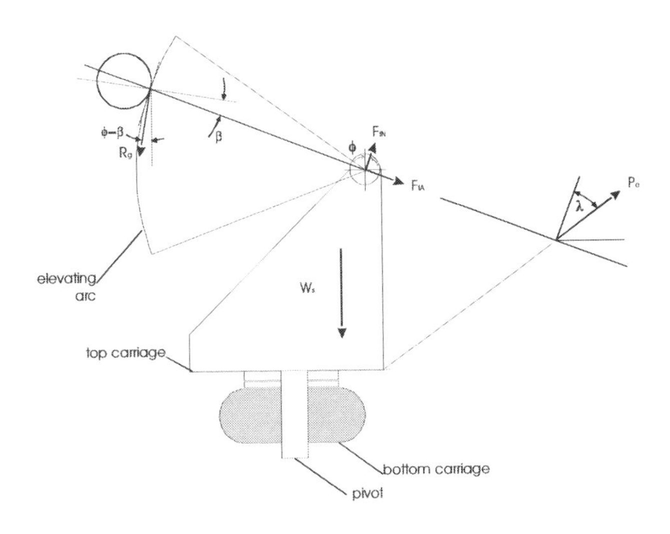

Fig 5.3.6: Loads on the bottom carriage

The balancing gear load is resolved as under:

$$H_e = P_e Cos[90 - (\varphi + \lambda)] = P_e Sin(\varphi + \lambda) \quad [5.3.18]$$

$$V_e = P_e Sin[90 - (\varphi + \lambda)] = P_e Cos(\varphi + \lambda) \quad [5.3.19]$$

The load on the elevating arc is resolved similarly:

$$H_g = R_g Sin(\phi - \beta) \quad [5.3.20]$$

$$V_g = R_g Cos(\phi - \beta) \quad [5.3.21]$$

The horizontal load on the bottom carriage which is transmitted from the traversing pivot to the bottom carriage, through the trails to the ground is obtained from the summation of the horizontal load components:

$$F_H = H_t + H_e - H_g \quad [5.3.22]$$

The vertical load is transmitted from the top carriage through the traverse bearing to the bottom carriage. Thereafter through the horizontal pin to the axle tree, the axles and finally through the wheels to the ground. Summing up the vertical components of forces on the top carriage, the vertical force on the bottom carriage is:

$$F_v = V_t + V_g - V_e + W_s \quad [5.3.23]$$

6

Elevating and Traversing Mechanisms

6.1: Gearing in General

Elevating and Traversing Mechanisms

Accurate aiming of the barrel in azimuth and in elevation is singularly important in the context of free flight projectiles. Aiming may be by direct laying on the target, as in the case of anti tank engagement, indirect laying as in the case of long range artillery or in conjunction with fire control units, as is the case with anti aircraft artillery. Aiming involves the speedy movement of heavy masses in vertical and horizontal planes. This calls for a transmission system, which responds quickly and sensitively to the signal from the source, which may be human or electronic.

The angle of elevation is imparted to the barrel by rocking it about its trunnions. The trunnions, because of the need for axial movement of the gun during recoil, have been, after the early cannon, universally shifted to the cradle. The transmission train, which effects this movement, beginning at the hand wheel and ending in the elevation, or depression, of the cradle with the barrel, is the elevating mechanism.

Similarly, the traversing mechanism serves the purpose of aligning the barrel in azimuth, in accordance with the requirement.

Both mechanisms are essentially gear trains or hydraulic or pneumatic linkages between the power source and the mass to be moved. To hold the barrel in the intended position, a component that ensures non-reversibility, such as self-locking worms or mechanical brakes, is incorporated in the train.

Choice of a power source follows from the tactical employment of the weapon, implying its desired speed of reaction and the mass of the components to be moved. Basically power sources are manual in the case of high torques where speed is not vital. Mechanical or electrically powered transmissions are employed in systems where high speed at high torques and for continuous periods, is called for. To cater for contingencies of power failure, powered systems invariably have manual back up.

General Requirements of Elevating and Traversing Gearing

- Access to the gearing, whether manual or power assisted, must be conveniently placed for easy access during operation as well as for maintenance.
- It must be easy to operate and demand only reasonable effort.
- Gearing, especially since associated with fire control systems, must afford the requisite degree of accuracy.
- It must incorporate the correct sense of direction.
- It should be non-reversible when required.
- Gearing should suffer from minimum backlash.

- For manually operated systems, a rough counter must be incorporated, by careful choice of the gear ratio. This will give the operator a fair idea of the magnitude of traverse and elevation per turn or half turn of the hand wheel.

Types of Bearings

Bearings are an indispensable feature of transmission systems. In the context of elevating and traversing gear trains, as will be seen, the bearings are invariably subject to high impulse loads, during firing. This aspect influences the selection of material and design of the bearings. Bearings may be of dissimilar metals, steel on steel or of ball or roller type.

Types of Gears

Elevating and traversing mechanisms incorporate different types of gear pairs, depending on design and layout. Gear pairs may be nut and screw, worm and segment or of arc and pinion type. A train will normally comprise of a combination of gear pairs. Bevel gears may be employed when a change of direction is implied.

Traversing Mechanisms

A traversing mechanism moves the top carriage in the horizontal plane, by transmitting motion to the top carriage to position it relative to the basic structure be it the bottom carriage or the tank hull. As earlier stated, traversing mechanisms may be manually operated, hydraulically or electrically driven. The common types of traversing mechanisms are:

- Worm and segment.
- Nut and screw.
- Arc and pinion.

Elevating Mechanisms

The elevating mass consists of the recoiling parts and the cradle. When the gun is fired the turning moment on the recoiling mass exerts stresses on the elevating gear. Therefore the ideal location of the elevating gear is in the same vertical plane as the axis of the bore. If it is mounted to one side, it is duplicated at an equal distance on the other side. During transportation, the cradle is locked to the carriage to prevent stress on the elevating gear by means of a cradle clamp.

Elevating and Traversing Stops

For reasons of stability, the gun should not fire out of the arc of traverse in which it has been designed to be stable. Also, to prevent fouling of the trails by the breech, traversing stops are provided with the traversing mechanism.

Stops are necessary to prevent the breech striking the ground during recoil, when firing at high angles of elevation. The gun tends to instability at larger angles of depression. This is taken into account at the design stage of the equipment and the angle of depression is restricted to desired limits by the incorporation of stops.

In the case of manually operated mechanisms, the stops consist of lugs on the movable member, mating with teeth or protrusions on the static member. In the case of powered mechanisms, where accelerations are of a high order, friction brakes or buffers are used to preempt impact between mobile and static components.

6.2: Elevating Gears Loads & Torques

Static Loads on Elevating Gears

Loads on the elevating gears that are present due to weight components of the elevating parts, loads on account of balancing and bearing friction loads are termed as static loads.

Moment after Balancing

With reference to Fig 6.2.1, the moment after balancing is given by the difference between the weight moment of the elevating parts and the balancing moment of the balancing gear about the axis of the trunnions:

$$T_e = M_{ob} - M_b \qquad [6.2.1]$$

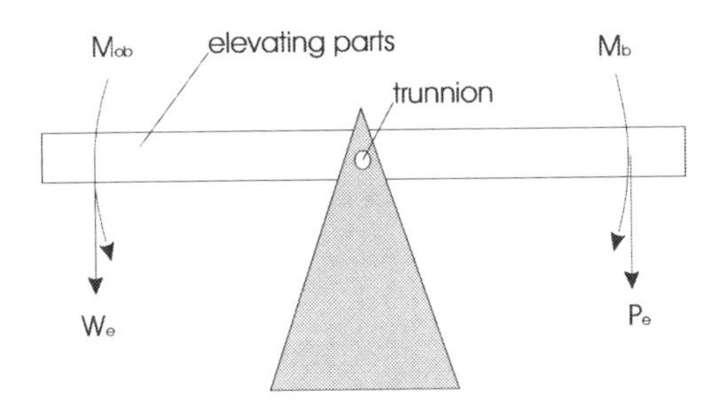

Fig 6.2.1: Weight moment after balancing

W_e: elevating weight
P_e: balancing gear force
M_{ob}: weight moment
M_b: balancing gear moment

Static Friction Force Moment in the Bearings:

The static friction moment in the trunnion bearings, illustrated in Fig 6.2.2, is given by:

$$T_b = \mu F_t r_b \quad [6.2.2]$$

F_t: load on trunnion bearing
r_b: bearing radius
μ:: coefficient of friction.

As recoil braking force increases, the load on trunnion bearing F_t increases hence the static friction moment T_b increases with increase in recoil force.

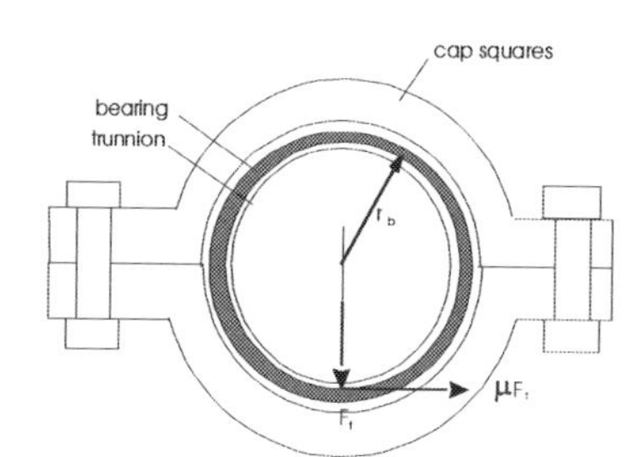

Fig 6.2.2: Static friction moment in trunnion bearings

Dynamic Loads on Elevating Gears

Loads on the elevating gearing that occur on firing are termed as dynamic loads.

Firing Couple

Due to the eccentricity of the loads applied about the trunnion axis during firing, a couple called the firing couple is generated at the elevating arc. This is depicted in Fig 6.2.3.

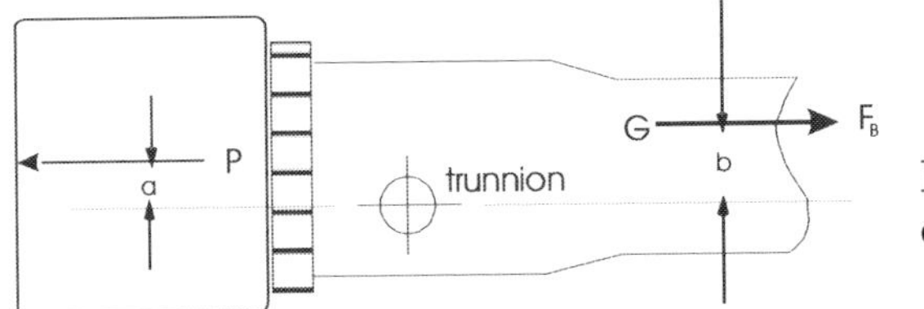

Fig 6.2.3: Firing couple on elevating mechanism

The firing couple which acts on the elevating gearing is given by:

$T_f = aP - bF_B$ .... [6.2.3]

P: the gas force
F_B: braking force of recoil system
G: centre of gravity of recoiling parts

While it is ideal to have the axis of the bore and the centre of gravity of the elevating parts on the trunnion axis, practically it is not possible due to space and structural constraints. Non-homogeneity of material and manufacturing tolerances also contribute to the firing torque. The effort is to minimize the firing torque by careful design. The firing torque as can be seen from Equation [6.2.3] is also dependent on the gas pressure, which varies considerably during the firing cycle. The duration of gas pressure influence is small compared to the recoil time. Two choices emerge; firstly it is possible to locate the centre of gravity of the recoiling parts in the plane of the trunnion axis. This will result in a zero firing torque except for the period of gas influence. Secondly if the bore axis and the trunnion axis are coplanar, the much larger torque due to gas pressure vanishes and the lesser torque due to recoil force remains.

Elevating Torque

The elevating torque is the torque required to accelerate the elevating weight. Given by:

$T_{aE} = I_E \alpha_E$ .... [6.2.4]

I_E: mass moment of inertia of the elevating parts about the trunnion axis
α: angular acceleration of the elevating parts

The net torque at the elevating arc is given by:

$$T_E = T_{\alpha E} + T_b + T_f + T_e \quad [6.2.5]$$

Manual Elevation

For a gun, which is elevated manually and does not fire while being elevated, the acceleration of the elevating parts is nonexistent and the firing couple is also zero. The relevant expression is:

$$T_E = T_b + T_e \quad [6.2.6]$$

Power Elevation

For a gun which is elevated by power elevation and which does not fire during the process of elevation:

$$T_E = T_{\alpha E} + T_b + T_e \quad [6.2.7]$$

Gearing

The purpose of the elevation gearing is to step down the large elevating torque required at the elevating arc to a lesser and manageable torque at the hand wheel, in the case of manual operation, or at the power source in the case of power driven elevating systems. By convention, the gear ratio of a gear train is taken as the ratio of the torque at the source, to the torque at the point of application, or the ratio of the speed at the input, of the train, to the speed at the output of the train, whichever is greater than 1. A basic elevating gear train is depicted in Fig 6.2.4 ahead.

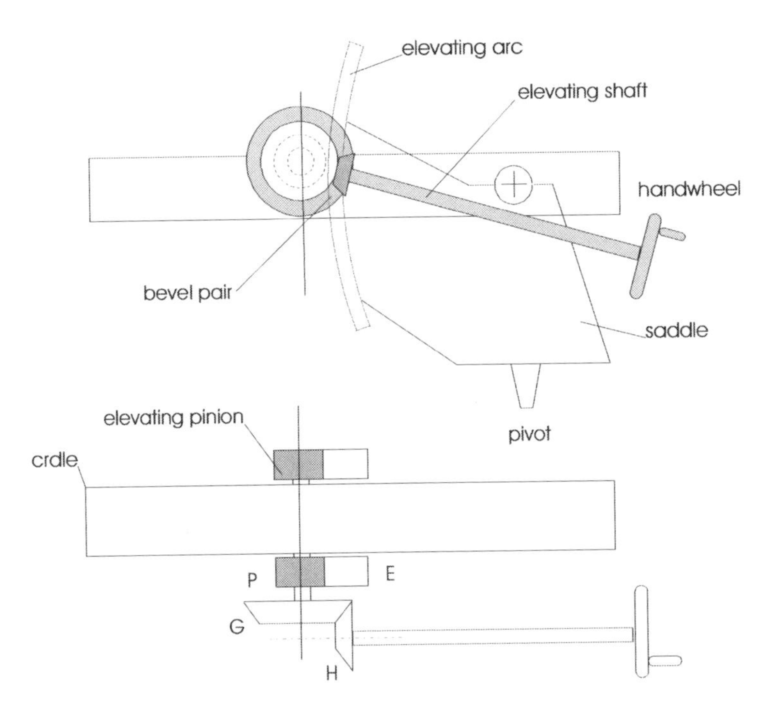

Fig 6.2.4: Simple manual elevating gear train

Manual Operation

The torque is traced down the train beginning at the elevating arc E to the hand wheel by converting it to gear tooth load and back to torque at each stage, applying the efficiency η at each meshing.

Gear tooth load between pinion P and elevating arc E is:

$$F_{PE} = \frac{T_E}{r_E}$$

r_E: radius of elevating arc

The torque in bevel gear G and pinion P becomes:

$$T_{GP} = \frac{1}{\eta} r_P F_{PE} = \frac{1}{\eta} T_E \frac{r_P}{r_E}$$

The gear tooth load between bevel gear H and bevel gear G is:

$$F_{HG} = \frac{T_{GP}}{r_G} = \frac{1}{\eta} T_E \frac{r_P}{r_G r_E}$$

r_P: radius of bevel gear G
r_G: radius of elevating pinion P

The torque at the hand wheel or power source required for elevation becomes:

$$T = \frac{1}{\eta} r_H F_{HG} = \frac{1}{\eta^2} T_E \frac{r_P r_H}{r_G r_E} = \frac{1}{\eta^2} \frac{T_E}{G.R}$$

G.R.: overall gear ratio.

Expressed generally:

$$T = \frac{1}{\eta^n} \frac{T_E}{G.R.} \qquad [6.2.8]$$

Or:

Torque ratio is equal to overall efficiency times gear ratio

T: torque at source or hand wheel
n: number of gear meshes

Efficiency of Gears

Spur and Bevel Gears

For spur and bevel gears, η = .98 to .99

Worm Gears

When a pinion is replaced by a worm whose efficiency is η_w:

$$T = \frac{1}{\eta_w \eta^{n-1}} \frac{T_E}{G.R.} \qquad [6.2.9]$$

The efficiency of a worm is given by:

$$\eta = \frac{Cos\beta - \mu Tan\lambda}{Cos\beta + \mu Cot\lambda} \qquad [6.2.10]$$

β: pressure angle
λ: lead angle
μ: coefficient of friction

When frictional losses in the thrust bearing of the worm are taken into consideration:

$$\eta = \frac{Cos\beta - \mu Tan\lambda}{Cos\beta\left(1 + \mu_b \frac{r_b}{r_w} Cot\lambda\right) + \mu Cot\lambda\left(1 - \mu_b \frac{r_b}{r_w} Tan\lambda\right)} \qquad [6.2.11]$$

r_b: effective bearing diameter
r_w: pitch diameter of worm
μ_b: friction coefficient of the bearing.

Power Operation

In the case of power driven elevation systems, the inertia of the elevating parts and that of the gear train determines the effort necessary for achieving the stipulated acceleration of the elevating parts.

Torque Required to Accelerate a Gear Train

Consider the general case of two shafts X and Y geared together. Let shaft X rotate with a speed of N_x and shaft Y with a speed of N_y, both speeds being in revolutions per second. Then the gear ratio between the two shafts is given by:

$$G = \frac{N_x}{N_y} \; or \; G = \frac{N_y}{N_x}$$

By convention, the gear ratio is taken as the magnitude of the speed ratio or the torque ratio, whichever is greater than 1.

We have $GN_x = N_y$ or $\omega_y = G\omega_x$

$\omega = 2\pi N$: angular velocity in radians per second

Therefore:

$$\frac{d\omega_y}{dt} = G\frac{d\omega_x}{dt} \; or \; \alpha_y = G\alpha_x$$

α : angular acceleration.

By Newton's second law of motion as applied to rotational bodies, torque is directly proportional to rate of change of momentum. Expressed mathematically:

$$T_a \propto \frac{dI\omega}{dt} \; or \; T_a = I\frac{d\varpi}{dt}$$

I: mass moment of inertia of the body, being constant.

Hence the torque required to accelerate a shaft by acceleration α is given by:

$$T_\alpha = I\alpha$$

Torque Transmission in a Basic Elevating Gear Train

A compound elevating gear train in its basic form is depicted ahead:

The torque required to accelerate each shaft S, A and E of the train individually is:

$T_{\alpha S} = I_s\alpha_s, T_{\alpha A} = I_A\alpha_A$ and $T_{\alpha E} = I_E\alpha_E$

I_E: mass moment of inertia of the elevating parts about the trunnion axis
I_S, I_A: mass moments of inertia of shafts S and A, respectively.
α_E: angular acceleration of the elevating parts
α_S: angular acceleration of shaft S
α_A: angular acceleration of shaft A

The torque to be applied to shaft S in order to provide a torque $T_{\alpha A}$ on shaft A is:

$$T_{SA} = G_{SA}T_{\alpha A}$$

Where the gear ratio between shafts S and A is $G_{SA} = \frac{r_G}{r_S}$

Similarly, the torque to be applied to shaft S in order to provide a torque $T_{\alpha E}$ on shaft E is:

$$T_{SE} = G_{SE}T_{\alpha E}$$

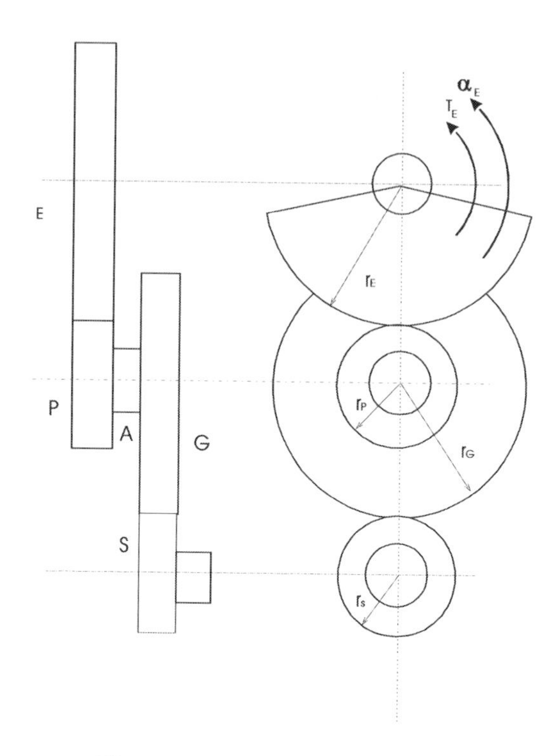

Fig 6.3.4: Elevating gear train

The gear ratio between shaft S and E is:

$$G_{SE} = \frac{r_G}{r_S}\frac{r_E}{r_P}$$

Taking efficiency into account, the net torque required to accelerate the elevating parts by an angular acceleration α is:

$$T_{\alpha E} = I_S\alpha_s + \frac{1}{\eta}\frac{r_G}{r_S}I_A\alpha_A + \frac{1}{\eta^2}\frac{r_G}{r_S}\frac{r_E}{r_P}I_E\alpha_E$$

$$\alpha_A = \frac{r_E}{r_P}\alpha_E; \quad \alpha_S = \frac{r_G}{r_S}\alpha_A = \frac{r_E}{r_P}\frac{r_G}{r_S}\alpha_E$$

$$T_{\alpha E} = \left(I_S \frac{r_E}{r_P}\frac{r_G}{r_S} + \frac{1}{\eta}\frac{r_G}{r_S}\frac{r_E}{r_P} I_A + \frac{1}{\eta^2}\frac{r_G}{r_S}\frac{r_E}{r_P} I_E \right)\alpha_E \quad \text{[6.2.12]}$$

From Equations [6.2.8] and [6.2.12], the net torque required at the input of the gear train or at the output of the hydraulic or electric drive motor is:

$$T_S = \frac{T_E}{\eta^n GR} + T_{\alpha E} \quad \text{[6.2.13]}$$

6.3: Traversing Gears Loads & Torques

Static Loads on the Traversing Mechanism

Static loads on the traversing mechanism are the loads, which are present due to weight components and friction forces arising in the traverse bearing.

Weight Moment about the Traversing Axis

When the equipment is on a slope and the center of gravity of the traversing parts does not coincide with the traversing axis about which the traversing takes place, a weight moment arises. This situation is depicted in Fig 6.3.1 ahead.

The weight moment about the traversing axis is given by:

$$M_T = W_t r Sin\theta Cos\psi \quad \text{[6.3.1]}$$

W_t: weight of traversing parts.
r: radius of traversing axis to CG of traversing parts.
θ: angle of inclination of the ground

ψ : angle between the bore axis and the line through the axis of traverse at right angles to the direction of the slope

On level terrain $\theta = 0$ which means $M_T = 0$

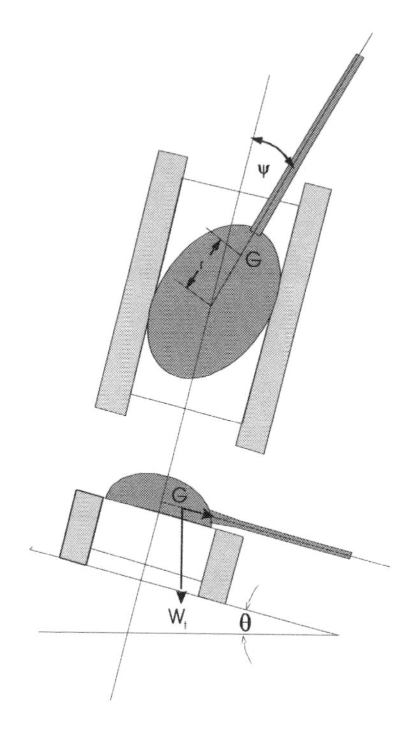

Fig 6.3.1: Weight moment due to ground inclination and offset of center of gravity of traversing parts from traversing axis

Torque due to Frictional Resistance in Traverse Bearing

Consider the elemental ring on the bearing face as in Fig 6.3.2. The normal force acting on this ring is:

$$dF_n = 2\pi p_n r dr$$

The total normal force acting on the bearing surface is:

$$F_n = \int_{r_i}^{r_o} 2\pi p_n r dr$$

Or:

$$F_n = \pi p_n \left(r_o^2 - r_i^2\right) \qquad [6.3.2]$$

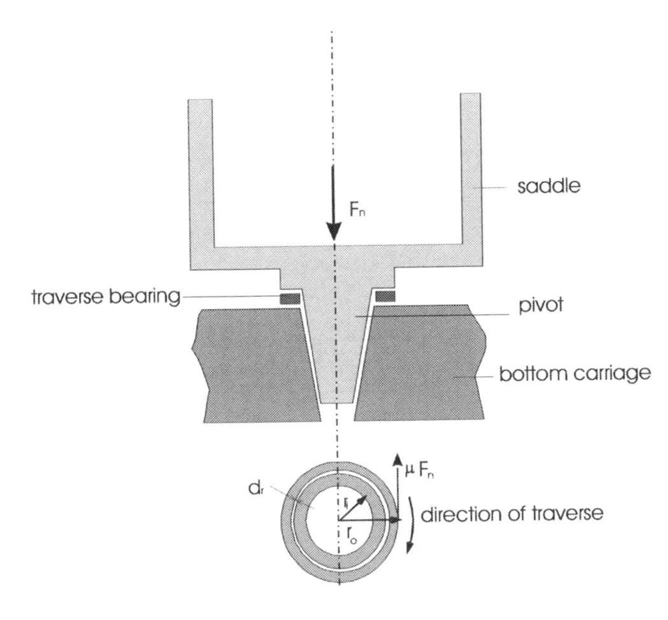

Fig 6.3.2: Moment due to bearing friction

p_n: normal pressure on traversing bearing
r: radius of elemental area
dr: width of elemental area
r_i: inner radius of bearing
r_o: outer radius of bearing

Friction torque on the element is:

$$dT_b = 2\pi\mu p_n r^2 dr$$

Net friction torque on the bearing is:

$$T_b = \int_{r_i}^{r_o} 2\pi\mu p_n r^2 dr = \frac{2}{3}\pi p_n \mu\left(r_o^3 - r_i^3\right)$$

Multiplying and dividing both sides by $\left(r_o^2 - r_i^2\right)$ and applying Equation [6.3.2]:

$$T_b = \frac{2}{3}\mu F_n \frac{\left(r_o^3 - r_i^3\right)}{\left(r_o^2 - r_i^2\right)} \qquad [6.3.3]$$

Dynamic Load on Traversing Mechanism

The dynamic load on the traversing mechanism is the load that occurs during firing.

Firing Couple due to the Eccentricity of the Recoiling Parts about the Traversing Axis

The line of action of the recoil braking force may not pass through the axis of traverse. Due to this eccentricity about the traversing axis, a couple as illustrated in Fig 6.3.3, is generated during firing, this is given by:

$$T_f = [aP - bF_B]Cos\phi \qquad [6.3.4]$$

F_B: braking force of recoil system
P: propellant gas force
ϕ : angle of elevation

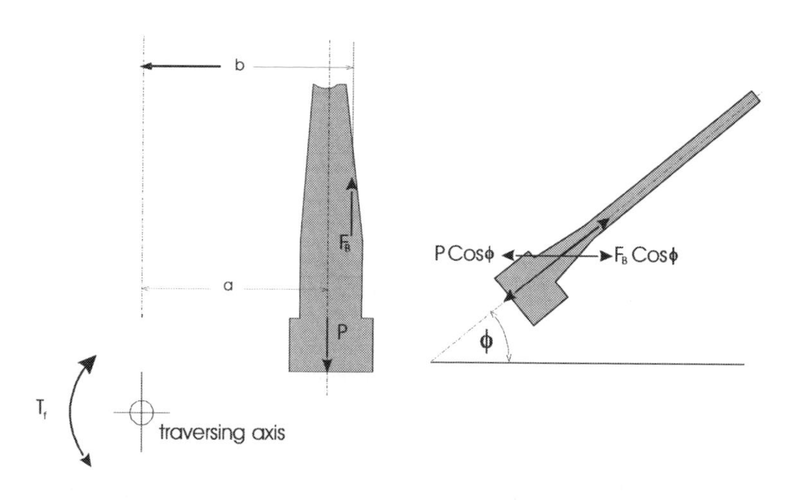

Fig 6.3.3: Firing couple on traversing mechanism

Torque Required to Accelerate the Traversing Parts

The fourth component of torque at the traversing gear is the torque required to accelerate the traversing parts, given by:

$$T_{aT} = I_T \alpha_T \qquad [6.3.4a]$$

I_T: mass moment of inertia of the traversing parts about the traversing axis
α_T: maximum desired acceleration of the traversing parts.

Maximum Torque at the Traversing Gear

Maximum torque at the traversing gear is given by:

$$T_T = M_T + T_b + T_f + T_{aT} \qquad [6.3.5]$$

Maximum Torque at the Traversing Gear under Different Conditions

When a gun on level ground is traversed manually and does not fire on the traverse:

$$T_T = T_b \qquad [6.3.6]$$

In the condition that the same gun is on an incline:

$$T_T = M_T + T_b \qquad [6.3.7]$$

For a gun, with power traverse, on level ground, if the gun does not fire while traversing:

$$T_T = T_b + T_{aT} \qquad [6.3.8]$$

For the same conditions but when the gun fires during traverse:

$$T_T = T_b + T_f + T_{dT} \qquad [6.3.9]$$

For the same gun which is on a slope and does not fire while traversing:

$$T_T = M_T + T_b + T_{dT} \qquad [6.3.10]$$

For the same gun which fires while traversing:

$$T_T = M_T + T_b + T_f + T_{dT} \qquad [6.4.11]$$

Torque at the Source

Torque at the source is calculated using the same technique, as for elevating mechanisms elucidated in the previous section.

Gear Train Ratio: AFV Mounted Guns

The choice of gear ratio depends on the nature of elevation required at the gun and the source of power at the actuating end. For manual elevation, average human effort is the criteria, whereas for power driven systems, speed is the determining factor. The tactical role of the gun determines the elevating and traversing accelerations. In the case of artillery, acceleration is not critical but ease of manual operation is important. In the case of tanks and anti aircraft cannon, which engage moving targets, high acceleration is vital for effective target engagement. The angular acceleration of such weapons is in the region of 0.5 radians/s.

Mobile Mountings

Mobile mountings or SP mountings have the advantages of greater stability due to heavier weight and low center of gravity. The firing stresses are transmitted directly to the ground through the tracks, which afford a larger contact area, resulting in reduced stresses. Hence there is less danger of damage to the axles or suspension system. The platform is more stable. This is advantageous from the accuracy and consistency point of view. High rates of elevation and traverse are possible. Power take off from the prime mover may be used to automate various functions. However SP mountings are complex and costly to design, manufacture and maintain.

7

Balancing

7.1: Balancing in General

Balancing

In rear trunnion weapons, the preponderance of weight is towards the muzzle. Therefore a means of correcting the out of balance moment is incorporated between the cradle and the saddle to permit easy elevation of the recoiling parts to any desired angle of elevation. As a natural but necessary consequence, controlled depression of the recoiling parts is achieved. The two methods of achieving this requirement are by balancing gears also called equilibrators or by balancing weights, sometimes by a combination of both.

Elevating Weight

The elevating weight consists of the weights of the barrel, the cradle and the recoil system which act forward of the trunnions.

Balancing Weights

Balancing or counter weights may weight added to the breech end of a rear trunnion gun to balance the weight of the gun forward of the trunnions. Balancing weights may be used on static equipment where the overall weight of the equipment is not limited by mobility considerations. The weight of the breech assembly also contributes substantially towards balancing. A disadvantage of a balancing weight is that it adds to the weight of the recoiling parts, increasing the magnitude of the recoil energy which has to be dissipated.

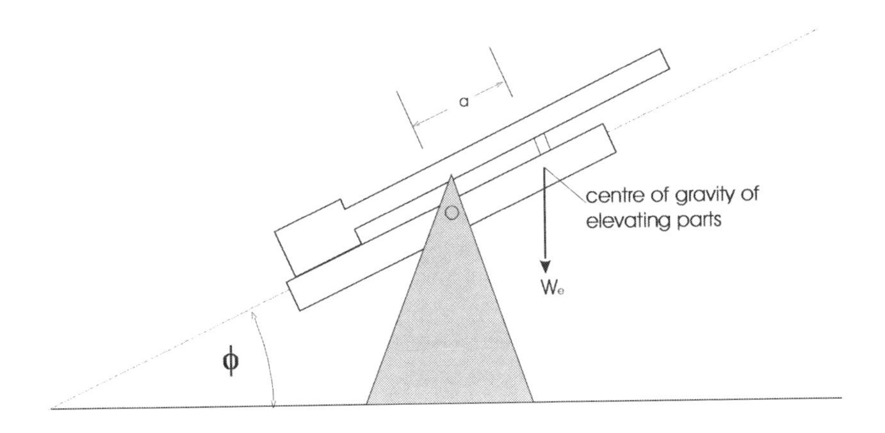

Fig 7.1.1: Out of balance moment

The assistance, or the balancing moment, provided by an balancing gear or balancing gear to overcome the moment due to the elevating weight varies with the Cosine of the angle of elevation and has to be equal and opposite to the out of balance moment. With reference to Fig 7.1.1:

$$M_{ob} = W_e a Cos\phi$$

M_{ob}: out of balance moment
a: moment arm
W_e: weight of elevating parts
ϕ : angle of elevation
G: centre of gravity of elevating parts

Types of Balancing Gears

Balancing gears may be either of the tension or compression type depending on whether they exert a pull or a push force to achieve the balancing moment. Operation wise, balancing gears may be based on metallic springs or of hydro-pneumatic type. Metal spring type gears are simple and reliable but their energy capacity is dependant on their physical size. Spring type balancing gears will function in compression even if the spring is fractured. Hydro-pneumatic balancing gears can store more energy in a smaller volume, but are characterized by a non-linear relationship between force and piston displacement. They also suffer from chronic sealing problems. Hydro-pneumatic systems are more adaptable because of flexible pressure lines, however in fighting vehicles, high-pressure fluid lines are a source of danger as the escape of high pressure fluid in the confined space of an AFV fighting compartment can have dangerous consequences if a rupture of the pipes occur. Hydro-pneumatic systems employ nitrogen or air for the compressible medium and oil for pressure transmission.

7.2: Tension Type Balancing Gears

Tension Type Balancing Gears

In the tension type balancing gears, the spring itself may be in compression or in tension but the linkage is in tension in both cases. Figs 7.2.2 and 7.2.3 show the simplified arrangements.

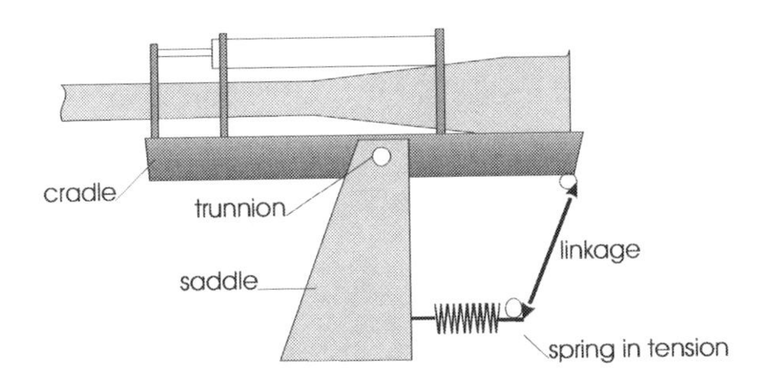

Fig 7.2.2: Spring in tension balancing gear

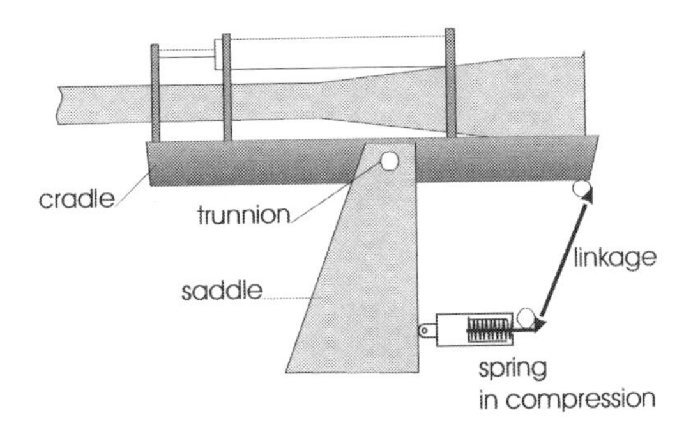

Fig 7.2.3: Spring in compression balancing gear

Springs in compression have advantages over springs in tension. Springs in compression are simpler to house between pressure faces. A tension spring has to terminate at visible positions. If a tension spring breaks, the components it is attached to will be subject to sudden displacement. Also energetic fragments of the spring are likely to cause injury to personnel in the vicinity. Hence from the safety point of view alone, springs in tension are disadvantageous.

Theory of Tension Type Balancing Gears

Tension types of balancing gears exert a pull on the gun for elevation. If the effect of friction be neglected, tension type balancing gears, properly designed, provide perfect balance over a range of 1600 mils. Tension types of balancing gears are generally preferred over the compression type as it will be shown; they can be designed to provide exact balance at all angles of elevation.

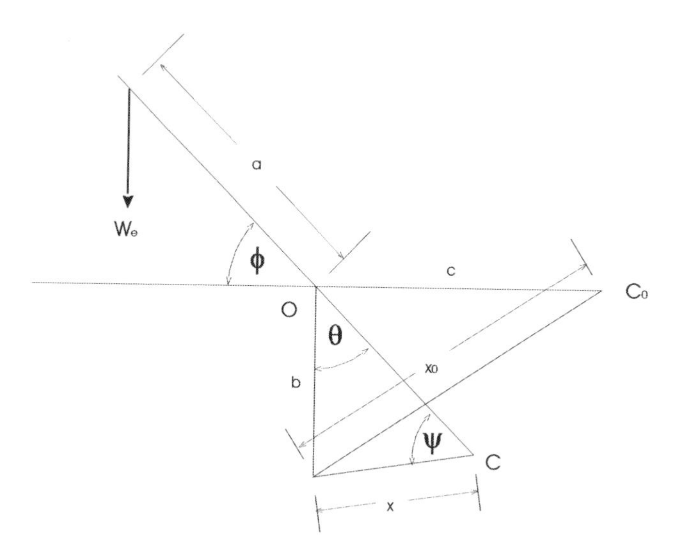

Fig 7.2.4: Force diagram of a tension type balancing gear

P_0: tension in the balancing gear spring at zero elevation.
P_e: tension in the balancing gear spring at elevation angle ϕ.

x_0: spring length at zero elevation
x: spring length at angle of elevation ϕ

Out of balance moment:

$$M_{ob} = W_e a \; Cos \; \phi \qquad [7.2.1]$$

Balancing moment:

$$M_b = P_e Sin \psi . c \qquad [7.2.2]$$

In triangle OBC:

$$\frac{b}{Sin \psi} = \frac{x}{Sin \theta} \text{ or } Sin \psi = \frac{b Sin \theta}{x}$$

Hence:

$$M_b = \frac{P_e b Sin \theta . c}{x} \qquad [7.2.3]$$

At $\phi = 0$, $M_{b0} = M_{ob0}$, $\theta = \theta_0 = \frac{\pi}{2}$ and $x = x_0$

$$M_{b0} = \frac{P_0 b Sin \theta_0 c}{x_0} \quad \text{and} \quad M_{ob0} = W_e a$$

For balance:

$$M_{ob0} = M_{b0}$$

Hence:

$$W_e a = \frac{P_0 b Sin \theta_0 c}{x_0}$$

Or:

$$P_0 = \frac{W_e a x_0}{bcSin\theta_0} \quad \text{.......} [7.2.4]$$

If the balancing gear spring stiffness is s, then:

$$P_e = P_0 - s[x_0 - x] = \frac{W_e a x_0}{bcSin\theta_0} - s[x_0 - x]$$

Substituting for P_e in Equation [7.2.3]:

$$M_b = \left[\frac{W_e a x_0}{bcSin\theta_0} - s[x_0 - x]\right]\frac{bcSin\theta}{x}$$

Adopting a spring stiffness:

$$s = \frac{W_e a}{bcSin\theta_0} \quad \text{.......} [7.2.5]$$

$$M_b = \left[\frac{W_e a x_0}{bcSin\theta_0} - \frac{W_e a}{bcSin\theta_0}[x_0 - x]\right]\frac{bcSin\theta}{x}$$

Or:

$$M_b = sSin\theta bc$$

Now:

$$\theta = \theta_0 - \phi$$

$$\therefore M_b = sbcSin(\theta_0 - \phi) = sbcCos\phi \quad \because \theta_0 = \frac{\pi}{2}$$

Or:

$$M_b = \frac{W_e a}{bcSin(\theta + \phi)} bcCos\phi$$

Hence for stability:

$$M_b = M_{ob} = W_e \, aCos\phi$$

The criteria for perfect balance from $\phi = 0$ to $\frac{\pi}{2}$:

- $\theta_0 = \frac{\pi}{2}$
- Spring stiffness $s = \frac{W_e a}{bc} \quad \because Sin\theta_0 = 1$
- Spring force at zero elevation: $P_0 = \frac{W_e a x_0}{bc}$

Example 7.2.1

Compute the spring stiffness, the initial spring force and plot the balancing and out of balance moment for a gun with a muzzle preponderance of 300 N. The center of gravity of the elevating weight is 0.75 m from the trunnions and on the axis of the bore. The gun is required to elevate from – 15 to + 45 degrees. The balancing gear force acts at a distance of 0.5 m from the trunnion axis towards the breech end. The point of attachment of the spring to the top carriage is 0.6 m directly under the trunnions.

Solution: Example 7.2.1

- Spring stiffness is calculated assuming the perfect balance condition.
- Initial spring length is calculated
- Initial spring force is calculated
- The balancing moment for angles of elevation of the gun is computed
- The balancing moment for angles of depression is computed.
- The balancing moments are plotted on the X axis against tangent of the angle of elevation on the Y axis.

Computer Programme 7.2.1

```
% programme to compute balancing moment: tension balancing gear

We=300 % elevating weight
a=.75 % moment arm elevating weight
b=.6 % trunnion to saddle attachment distance
c=.5 % trunnion to balancing gear attachment distance
s=We.*a./(b.*c) % spring stiffness
x0=sqrt(b.^2+c.^2) % initial spring length
P0=We.*a.*x0./(b.*c) % initial spring force
phie=linspace(0,pi/4,10) % elevation +
mbe=We.*cos(phie)*a % balancing moment in elevation
ye=tan(phie).*mbe
phid=linspace(-pi/12,0,4) % elevation-
mbd=we.*cos(phid)*a % balancing moment depression
yd=tan(phid).*mbd
axis square
plot(mbe,ye,mbd,yd)
```

Results: Example 7.2.1

Spring stiffness = 750 N/m		
Initial spring force = 585.77 N		
Elevation		Balancing moment Nm
radians	degrees	
-0.2618	-15	217.3333
-0.1745	-10	221.5817
-0.0873	-5	224.1438
0	0	225.0000
0.0873	5	224.1438
0.1745	10	221.5817
0.2618	15	217.3333
0.3491	20	211.4308
0.4363	25	203.9193
0.5236	30	194.8557
0.6109	35	184.3092
0.6981	40	172.3600
0.7854	45	159.0990

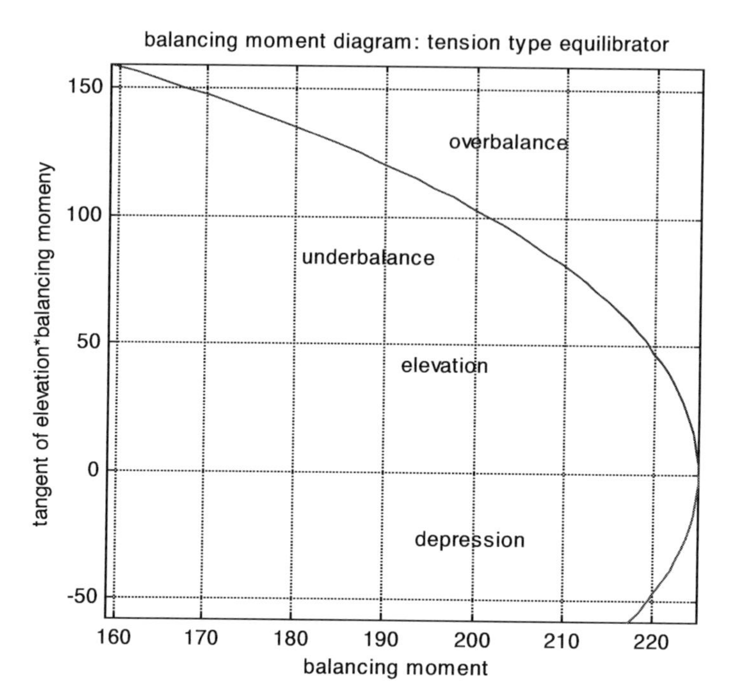

Fig 7.2.5: Balancing moment–angle of elevation diagram for tension type balancing gear of Example 7.2.1

7.3: Compression Type Balancing Gears

Limitations of a Compression Type Balancing Gear

This type of balancing gear cannot provide exact balance at all angles of elevation; hence it is designed to provide balance within a limited range of elevation angles associated with the equipment in question. Here the arm connecting the basic structure to the saddle is in compression. Balance to within 1% over a range of 0-90 degrees elevation is feasible if the dimensions are carefully chosen. The balancing gear exerts a push on the cradle during elevation.

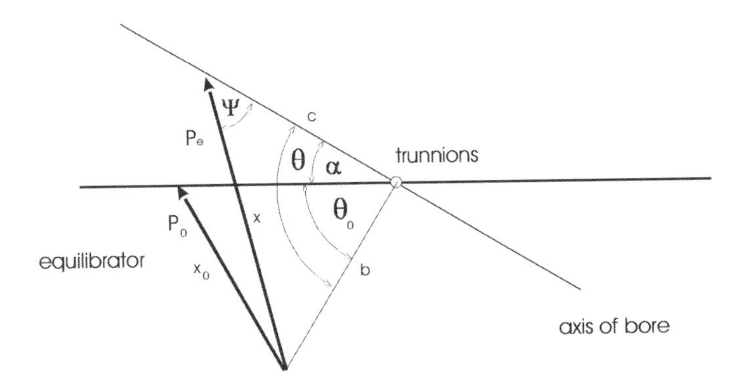

Fig 7.3.1: Force diagram of compression type balancing gear

Theory of Compression Type Balancing Gears

The balancing moment: $M_b = P_e c Sin\,\psi$.

Also from the triangle OAB:

$x^2 = b^2 + c^2 - 2bcCos\theta$ .... [7.3.1]

Since:

$$\frac{b}{\sin\psi} = \frac{x}{\sin\theta}, \sin\psi = \frac{b}{x}\sin\theta$$

Choosing $\theta = \theta_0$ at which $P_e = P_0$ and $x = x_0$:

$$M_{b0} = \frac{P_0 bc\sin\theta_0}{x_0}$$

Hence P_e for any other value of θ is P_0 plus the change in value of P_0 due to the change in value of θ from θ_0 to θ.

$$P_e = P_0 - s(x - x_0) \qquad [7.3.2]$$

s: spring stiffness
$x - x_0$: change in spring length

The balancing moment is:

$$M_b = [P_0 - s(x - x_0)]\frac{bc \sin\theta}{x} \qquad [7.3.3]$$

Hence the relationship between M_b and θ can be plotted against the standard out of balance moment curve $M_{ob} = W_e a Cos\phi$, where ϕ is the angle of elevation of the gun and a is the distance of the center of gravity of the elevating weight from the trunnion axis.

Example 7.3.1

The elevating weight of a gun is 1440N. With reference to Fig 7.3.1, a is 2m, b is 1m, and c is 5m. The initial spring force of the compression type helical spring balancing gear is 3126 N. θ_0 is taken as 30 degrees. Compute and plot the balancing and out of balance moments for elevation angles from 0 to 90 degrees in steps of 10 degrees.

Solution: Example 7.3.1

- Initial spring length is computed from the geometry of the figure using Equation [7.3.1].
- The spring length is computed for different values of angle of elevation from 0 to 90° using the same equation.
- The spring force is computed for different values of spring length using Equation [7.3.2].
- The out of balance moment is computed using Equation [7.2.1]
- The balancing moment is computed using Equation [7.3.3]

- The out of balance and the balancing moments are plotted against different values of elevation from 0 to 90°.

Computer Programme 7.3.1

```
% programme to compute o/b and balancing moments: compression type
spring balancing gear

We=1440.0 % elevating weight
a=2
b=1
c=1.5
P0=3126.0 % initial spring force N
s=1980.0 % spring stiffness N/m
theta0=30/180*pi % initial theta
x0=sqrt(b.^2+c.^2-2.*b.*c.*cos(theta0)) % initial spring length
psi=linspace(0,pi/2,10) % elevation angle
thetae=psi+theta0
x=sqrt(b.^2+c.^2-2.*b.*c.*cos(thetae)) % spring length
Pe=P0-s.*(x-x0)
Mob=We.*a.*cos(psi) % out of balance moment
Mb=Pe.*b.*c./x.*sin(thetae) % balancing moment
plot(psi,Mob,psi,Mb)
```

Results: Example 7.3.1

Elevation radians	Spring length m	Spring force N	O/b moment Nm	Bal moment Nm
0	0.8074	3.1260	2.8800	2.9037
0.1745	0.9756	2.7929	2.8362	2.7601
0.3491	1.1496	2.4484	2.7063	2.4472
0.5236	1.3229	2.1054	2.4942	2.0675
0.6981	1.4913	1.7719	2.2062	1.6748
0.8727	1.6520	1.4538	1.8512	1.3000
1.0472	1.8028	1.1552	1.4400	0.9612
1.2217	1.9419	0.8797	0.9850	0.6692
1.3963	2.0679	0.6303	0.5001	0.4296
1.5708	2.1794	0.4094	0.0000	0.2440

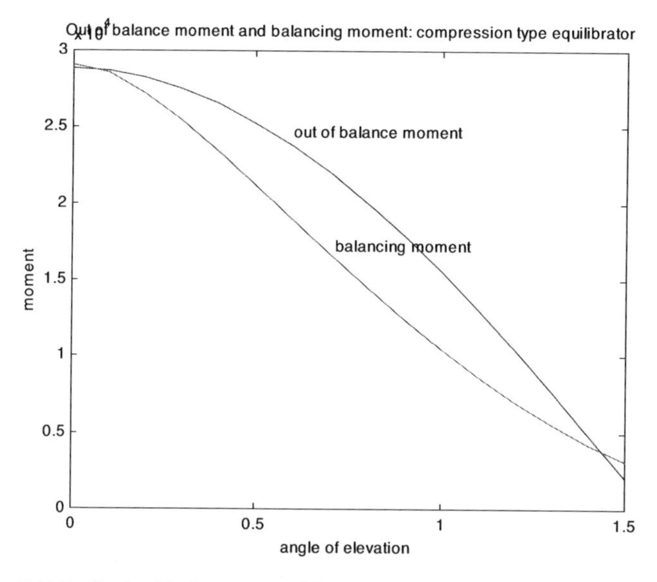

Fig 7.3.2: Out of balance and balancing moments: compression type balancing gear

Advantages and Disadvantages of Rear Trunnions with Balancing Gears

Rear trunnions facilitate easy loading, as the breech is low. A lower trunnion axis results in better stability. The carriage becomes smaller and lighter. In vehicles with fighting compartments more turret space is available. Wider traverse is possible and the overall silhouette of the weapon is reduced in height. The accompanying disadvantages of balancing gears is that the complexity of the system increases, the balancing gear, if pneumatic, becomes dependant on temperature variation and balancing is disturbed when the gun is on an incline as the initial conditions for which the balancing gear was designed are disturbed.

APPENDICES

Useful Conversion Tables

Length

	m	in	ft	yd
m	1	39.37	3.281	1.094
in	$2.540.10^{-2}$	1	$8.333.10^{-2}$	$2.778.10^{2}$
Ft	$3.048.10^{-1}$	12	1	$3.333.10^{-1}$
Yd	$9.144.10^{-1}$	36	3	1

Area

	m^2	In^2	ft^2	yd^2
m^2	1	$1.550.10^{3}$	$1.076.10^{1}$	1.196
in^2	$6.452.10^{-4}$	1	$6.944.10^{-3}$	$7.716.10^{-4}$
ft^2	$9.290.10^{-2}$	$1.440.10^{2}$	1	$1.111.10^{-4}$
yd^2	$8.361.10^{-1}$	$1.296.10^{3}$	9	1

Mass

	kg	lb	ton(UK)	ton(US)	grain
kg	1	2.205	$9.842.10^{-4}$	$1.102.10^{-3}$	$1.543.10^{4}$
lb	$4.535.10^{-1}$	1	$4.464.10^{-4}$	$5.0.10^{-4}$	$7.0.10^{3}$
ton (UK)	$1.016.10^{3}$	$2.240.10^{3}$	1	1.120	$1.567.10^{7}$
ton (US)	$9.072.10^{2}$	$2.0.10^{3}$	$8.929.10^{-1}$	1	$1.40.10^{7}$
grain	$6.480.10^{-5}$	$1.428.10^{-4}$	$6.378.10^{-8}$	$7.141.10^{-8}$	1

Volume

	m^3	litre	in^3	ft^3	yd^3
m	1	10^{3}	$6.102.10^{4}$	35.31	1.308
litre	10^{-3}	1	61.02	$3.531.10^{-2}$	$1.308.10^{-3}$
in^3	$1.639\text{-}10^{-5}$	$1.639.10^{-2}$	1	$5.787.10^{4}$	$2.143.10^{-5}$
ft^3	$2.832.10^{-2}$	28.32	$1.728.10^{3}$	1	$3.704.10^{-2}$
yd^3	$7.646.10^{-1}$	$7.646.10^{2}$	$4.666.10^{4}$	27	1

Angular Velocity

	rad/s	rad/min	degree/s	rps	rpm
rad/s	1	60	57.30	$1.592.10^{-1}$	9.549
rad/min	$1.667.10^{-2}$	1	$9.550.10^{-1}$	$2.653.10^{-3}$	$1.592.10^{-1}$
degree/s	$1.745.10^{-2}$	1.047	1	$2.778.10^{-3}$	$1.667.10^{-1}$
rps	6.283	$3.770.10^{2}$	$3.60.10^{2}$	1	60
rpm	$1.047.10^{-1}$	6.283	60	$1.667.10^{-2}$	1

Velocity

	m/s	m/min	km/h	Ft/s	Ft/min	mile/h
m/s	1	60	3.6	3.281	$1.969.10^{2}$	2.237
m/min	$1.667.10^{-2}$	1	6.10^{-2}	$5.468.10^{-2}$	3.281	$3.728.10^{-2}$
km/h	$2.778.10^{-1}$	16.67	1	$9.113.10^{-1}$	54.68	$6.214.10^{-1}$
ft/s	$3.048.10^{-1}$	18.29	1.097	1	60	$6.818.10^{-1}$
ft/min	$5.080.10^{-3}$	$3.048.10^{-1}$	$1.829.10^{-2}$	$1.667.10^{-2}$	1	$1.136.10^{-2}$
mile/h	$4.470.10^{-1}$	$2.682.10^{1}$	1.609	1.467	$8.80.10^{4}$	1

Density

	kg/m^3	ib/ft^3	lb/in^3
kg/m3	1	$6.243.10^{-2}$	$3.613.10^{-5}$
lb/ft^3	16.02	1	$5.787.10^{-4}$
lb/in^3	$2.768.10^4$	$1.728.10^3$	1

Acceleration

	m/s^2	ft/s^2
m/s^2	1	3.281
ft/s^2	$3.048.10^{-1}$	1

Force

	N	kgf	dyne	lbf	tonf UK	tonf US	poundal
N	1	$1.020.10^{-1}$	$1.0.10^{5}$	$2.248.10^{-1}$	$1.004.10^{-4}$	$1.124.10^{-4}$	7.233
kgf	9.807	1	$9.807.10^{5}$	2.205	$9.842.10^{-4}$	$1.102.10^{-3}$	70.93
dyne	1.10^{-5}	$1.020.10^{-6}$	1	$2.248.10^{-6}$	$1.004.10^{-9}$	$1.124.10^{-9}$	$7.233.10^{-5}$
lbf	4.448	$4.536.10^{-1}$	$4.448.10^{5}$	1	$4.464.10^{-4}$	5.10^{-4}	32.17
tonf UK	$9.964.10^{3}$	$1.016.10^{3}$	$9.964.10^{8}$	$2.240.10^{3}$	1	1.120	$7.207.10^{4}$
tonf US	$8.896.10^{3}$	$9.075.10^{2}$	$8.893.10^{3}$	2.10^{3}	$8.932.10^{-1}$	1	$6.433.10^{4}$
pdl	$1.383.10^{-1}$	$1.410.10^{-2}$	$1.383.10^{4}$	$3.108.10^{2}$	$1.388.10^{-5}$	$1.554.10^{-5}$	1

Moment of Force

	Nm	kgfm,	lbf ft
Nm	1	1.020^{-1}	$7.376.10^{-1}$
kgfm	9.807	1	7.233
lbf ft	1.356	$1.382.10^{-1}$	1

Energy

	J	kwh	ft lbf
J	1	$2.778.10^{-7}$	$7.376.10^{-1}$
kwh	$3.6.10^{6}$	1	$2.655.10^{6}$
ft lbf	1.356	$3.766.10^{-7}$	1

Pressure

	Pa	bar	Kgf/cm²	atm	psi
Pa	1	1.10^{-5}	$1.020.10^{-5}$	$9.869.10^{-6}$	$1.450.10^{-4}$
bar	1.10^{5}	1	1.020	$9.869.10^{-1}$	14.51
kgf/cm²	$9.807.10^{4}$	$9.807.10^{-1}$	1	$9.678.10^{-1}$	14.22
atm	$1.013.10^{5}$	1.013	1.033	1	14.70
psi	$6.895.10^{3}$	$6.895.10^{-2}$	$7.031.10^{-2}$	$6.805.10^{-2}$	1

Multiples of Pa: GPa, MPa, KPa

Stress

	Pa	N/mm^2	N/cm^2	kgf/mm^2	kgf/mm^2	lbf/in^2	$tonf/in^2$ (UK)
Pa	1	1.10^{-6}	1.10^{-4}	$1.020.10^{-7}$	$1.020.10^{-5}$	$1.450.10^{-4}$	$6.475.10^{-8}$
N/mm^2	1.10^{6}	1	100	$1.020.10^{-1}$	10.20	145.00	$6.475.10^{-2}$
N/cm^2	1.10^{4}	1.10^{-2}	1	$1.020.10^{-3}$	$1.020.10^{-1}$	1.450	$6.475.10^{-4}$
kg/mm^2	$9.807.10^{6}$	9.807	$9.807.10^{2}$	1	100	1422.00	$6.350.10^{-1}$
kg/cm^2	$9.807.10^{4}$	$9.807.10^{-2}$	9.807	1.10^{-2}	1	14.22	$6.350.10^{-3}$
lbf/in^2	$6.895.10^{3}$	$6.895.10^{-3}$	$6.895.10^{-1}$	$7.031.10^{-4}$	$7.031.10^{-2}$	1	$4.464.10^{-4}$
$tonf/in^2$ (UK)	$1.544.10^{7}$	1.5.44	$1.544.10^{3}$	1.575	157.50	2240.00	1

Power

	w	HP	ft lbf/s
w	1	$1.341.10^{-3}$	$7.376.10^{-1}$
HP	$7.457.10^{2}$	1	$5.50.10^{2}$
ft lbf/s	1.356	$1.818.10^{-3}$	1

Moments of Inertia of Sections about the Neutral Axis as Shown

Description of Section	Moment of Inertia about the Neutral Axis
Rectangular lamina of breadth b and depth d 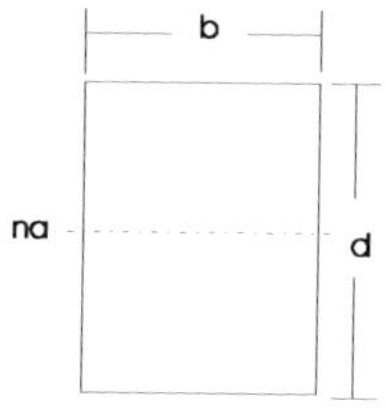 b na d	$\frac{bd^3}{12}$
Parallelogram of side a 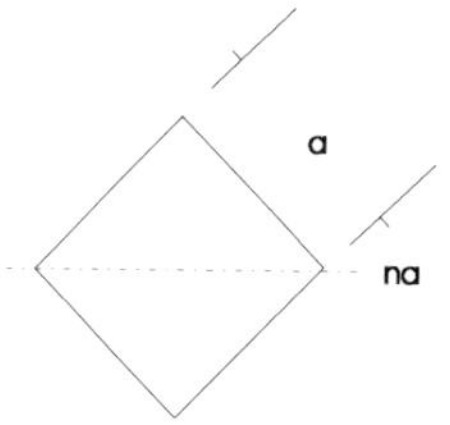 a na	$\frac{a^4}{12}$

Circular disc of diameter d 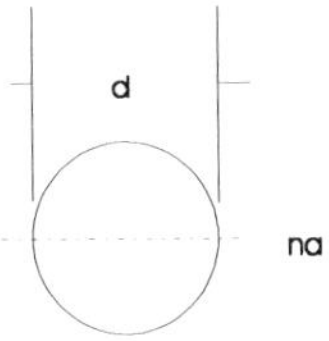 d na	$\frac{\pi}{64}d^4$
Annular disc of outer diameter D and inner diameter d 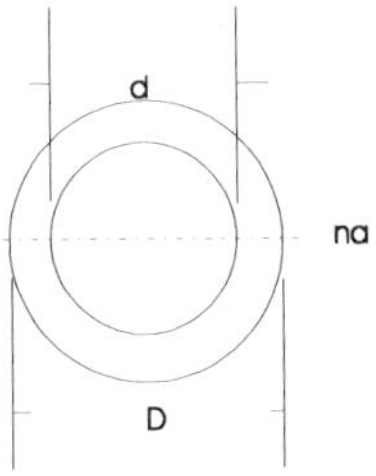 d na D	$\frac{\pi}{64}\left(D^4 - d^4\right)$
Elliptical disc 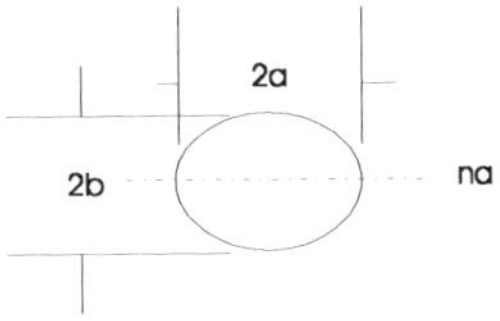 2a 2b na	$\frac{\pi}{4}ab^3$

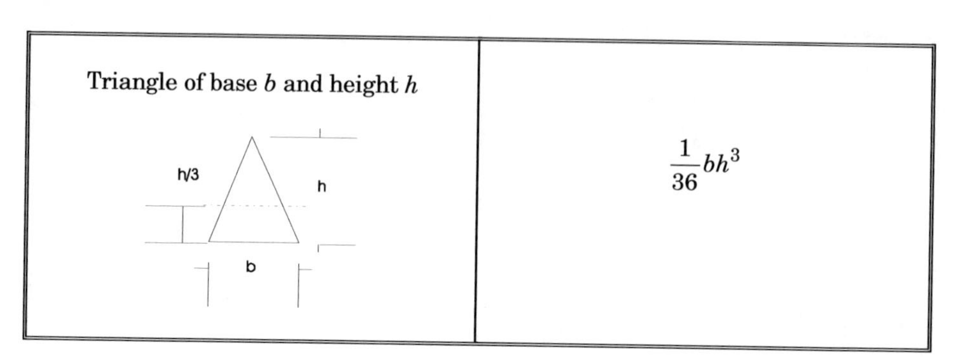

Triangle of base b and height h	$\frac{1}{36}bh^3$

Moments of Inertia of Solids

Moments of inertia of various solid bodies of uniform density and of mass m.

Description of Body	Moment of Inertia
Thin rod of length l • About the axis perpendicular to the rod through the center of mass, • About the axis perpendicular to the rod through one end.	$\frac{1}{12}ml^2$ $\frac{1}{3}ml^2$
Rectangular lamina with sides a, b • About an axis perpendicular to the plate through its center. • About an axis parallel to side b through its center.	$\frac{1}{12}m(a^2+b^2)$ $\frac{1}{12}ma^2$

Circular cylinder of radius r and length l • About axis of cylinder, • About axis through center of mass and perpendicular to cylindrical axis, • About axis coinciding with diameter at one end.	$\frac{1}{2}mr^2$ $\frac{1}{12}m(l^2+3r^2)$ $\frac{1}{12}m(4l^2+3r^2)$
Hollow circular cylinder of outer radius r_o, inner radius r_i and length l • About the axis of cylinder. • About the axis through center of mass and perpendicular to cylindrical axis. • About the axis coinciding with a diameter at one end.	$\frac{1}{2}m(r_0^2+r_i^2)$ $\frac{1}{12}m(3r_i^2+3r_0^2+l^2)$ $\frac{1}{12}m(3r_i^2+3r_0^2+4l^2)$

Circular plate of radius r • About axis perpendicular to plate through center. • About axis coinciding with a diameter.	$\frac{1}{2}mr^2$ $\frac{1}{4}mr^2$
Hollow circular plate or ring with outer radius r_0 and inner radius r_i • About axis perpendicular to plane of plate through center, • About axis coinciding with a diameter.	$\frac{1}{2}m(r_0^2 + r_i^2)$ $\frac{1}{4}m(r_0^2 + r_i^2)$
Thin circular ring of radius r • About axis perpendicular to plane of ring through center. • About axis coinciding with diameter.	mr^2 $\frac{1}{2}mr^2$
Sphere of radius r • About axis coinciding with a diameter, • About axis tangent to the surface.	$\frac{2}{5}mr^2$ $\frac{7}{5}mr^2$

Hollow sphere of outer radius r_o and inner radius r_i • About axis coinciding with a diameter. • About axis tangent to the surface.	$\frac{2}{5}m\left(r_o^5 - r_i^5\right)\left(r_o^3 - r_i^3\right)$ $\frac{2}{5}m\left(\frac{r_o^5 - r_i^5}{r_o^3 - r_i^3}\right) + mr_o^2$
Hollow spherical shell of radius r • about axis coinciding with a diameter, • About axis tangent to the surface.	$\frac{2}{3}mr^2$ $\frac{5}{3}mr^2$

Geometric Formulae

Triangle of altitude h and base b
Area $= bh = ab\ \sin\theta =$
$\sqrt{\left(s(s-a)(s-b)-(s-c)\right)}$
Where: $s = \dfrac{1}{2}(a+b+c) =$ semi perimeter

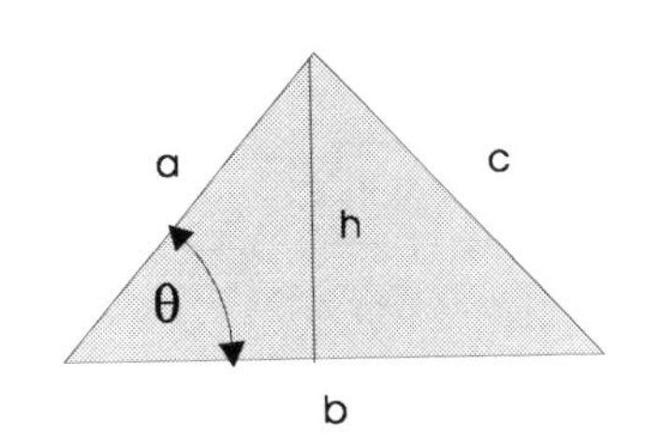

Sector of circle of radius r
Area $= \dfrac{1}{2}r^2\theta$, θ is in radians
Arc length $s = r\theta$

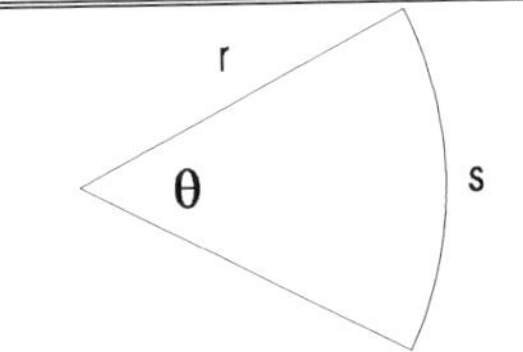

Segment of circle of radius r
Area $= \dfrac{1}{2}r^2(\theta - \sin\theta)$

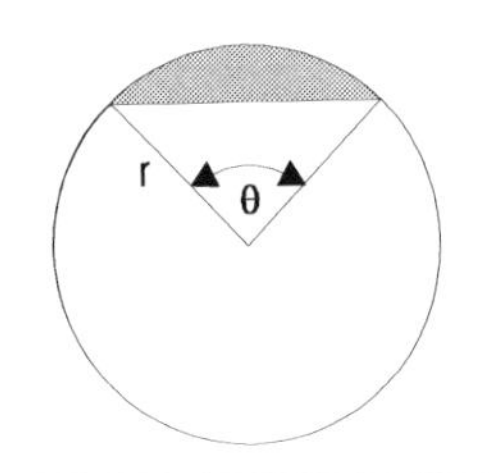

Ellipse of semi-major axis a and semi-minor axis b Area = πab Perimeter = $4a\int_0^{\pi/2}\sqrt{1-k^2\sin^2\theta\,d\theta}$ $=2\pi\sqrt{\frac{1}{2}(a^2+b^2)}$ Where $k=\sqrt{a^2-b^2/a.}$	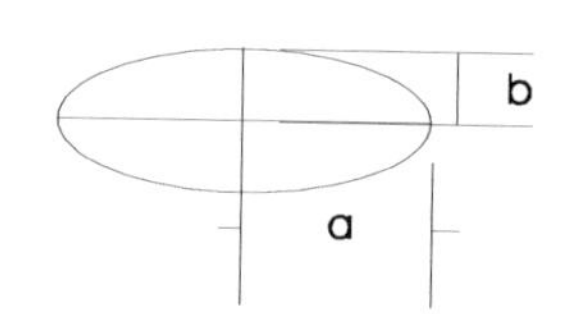
Segment of a parabola ABC Area = $\frac{2}{3}ab$ Arc length ABC = $\frac{1}{2}\sqrt{b+16a^2}+\frac{b^2}{8a}\ln\left(\frac{4a+\sqrt{b^2+16a^2}}{b}\right)$	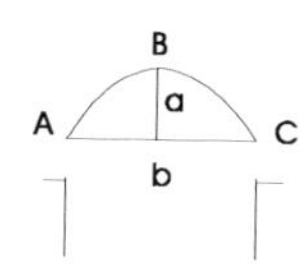
Right circular cone of radius r and height h Volume = $\frac{1}{3}\pi r^2h$ Lateral surface area = $\pi r\sqrt{r^2+h^2}=\pi rl$	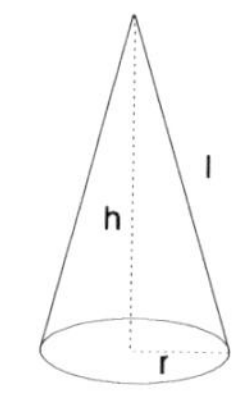
Spherical cap of radius r and height h Volume (shaded) = $\frac{1}{3}\pi h^2(3r-h)$ Surface area = $2\pi rh$	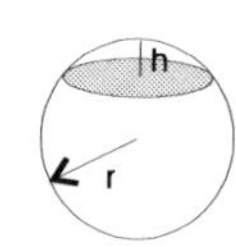

Frustum of right circular cone of radii r_1, r_2 and height h Volume = $\frac{1}{3}\pi h(r_1^2 + r_1 r_2 + r_2^2)$ Lateral surface area = $\pi(r_1 + r_2)\sqrt{h^2 + (r_2 - r_1)^2}$ $= \pi(r_1 + r_2)l$	r_1 h r_2
Spherical triangle of angles A, B, C on sphere of radius r Area of triangle $ABC = (A + B + C - \pi)r^2$	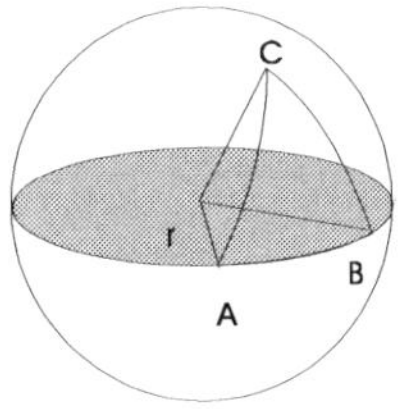 C r B A
Torus of inner radius r_i and outer radius r_o Volume = $\frac{1}{4}\pi^2(r_o + r_i)(r_o - r_i)^2$ Surface area = $\pi^2(r_o^2 - r_i^2)$	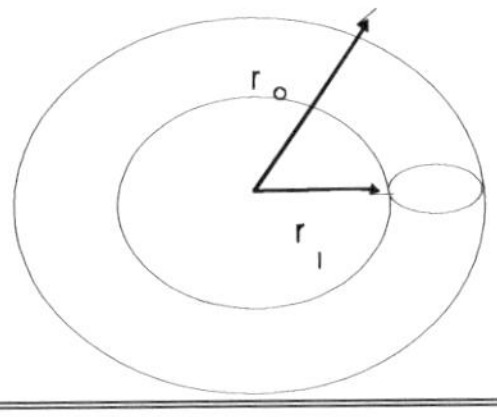 r_o r_i
Ellipsoid of semi-axes a,b,c Volume = $\frac{4}{3}\pi abc$	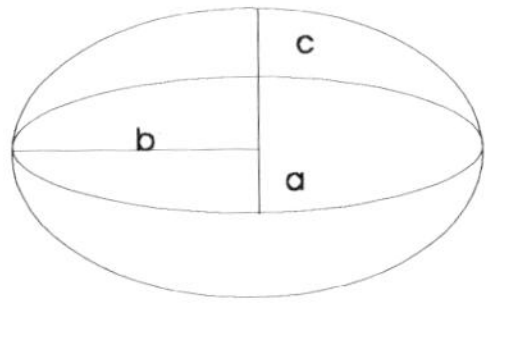 c b a

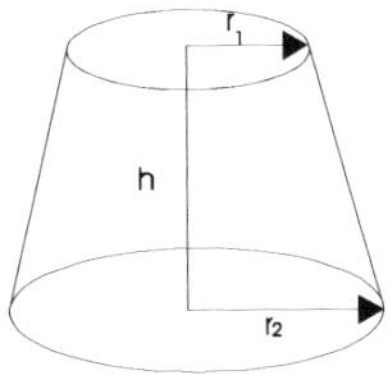

Paraboloid of revolution Volume = $\frac{1}{2}\pi b^2 a$	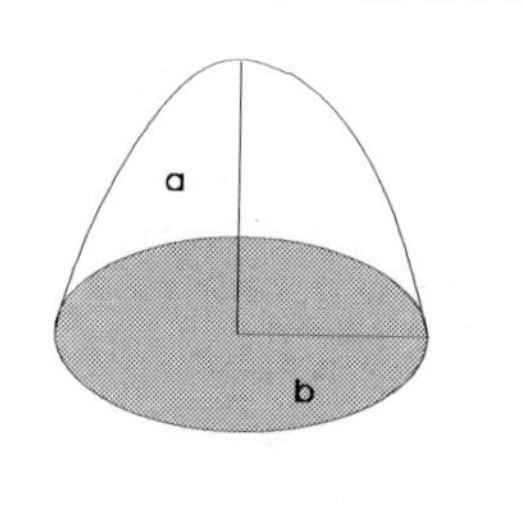

References
(In alphabetical order)

130mm gun Model 1946, Service Manual
76 mm Gun Model 1942, Service Manual
122mm Howitzer D 30, Service Manual
A Text Book of Applied Mechanics, 4th Edition 1983, S Ramamrutham, Dhanpat Rai & Sons
A Text book of Fluid Mechanics ,RS Khurmi ,15th edition,1989, S Chand & Co.
A Text Book of Strength of Materials, 6th Edition 1975, IB Prasad, Khanna Publishers
Advanced Fluid Mechanics, Volume 1, 1958, RC Binder, Prentice Hall Inc.
Applied Fluid Mechanics, 2nd Edition 1979, Robert L Mott, Charles E Merill Publishing Company
Elements of Strength of Materials, S Timoshenko & Gleason H MacCullough, 3rd Edition1949, D Van Nortrand Company Inc.
Foundations of Fluid Mechanics, SW Yuan, Prentice Hall of India, 1988.
Guns Mortars and Rockets Revised edition 1997,MP Manson, Brasseys (UK) Ltd
Machine Design An Integrated Approach, 1996, Robert L Norton, Prentice Hall
Machine Design, 13th Edition,1997, Pandya and Shah, Charotar Publishing House
Mechanics of Engineering Materials,2nd Edition 1997, PP Benham, RJ Crawford, CG Armstrong, Addison Wesley Longman Ltd.
Mechanics of Materials 3rd Edition 1997, EJ Hearn ,Butterworth Heinmann
Microsoft Excel in 24 Hours, Trudi Reisner, Techmedia
Oerlikon Pocket Book 2nd Edition 1981, Oerlikon-Buhrle AG
Schaums Outline Series Theory and Problems of Strength of Materials 2nd Edition,1972, William A Nash, McGraw Hill Inc.
Solving Problems in Fluid Mechanics, JF Douglas, Longman Scientific & Technical, 1989.

Strength of Materials, 1973, V Feodosyev, Mir Publishers
Tanks Design & Calculation, SS Bunov
Text Book of Service Ordnance 1923
The Student Edition of Matlab Version 4, Users Guide, The Mathworks Inc, Prentice Hall Inc
Theory of Elasticity, SP Timoshenko & JN Goodier, 3rd Edition 1970, McGraw Hill
Theory of Machines, 8th Edition, 1986, RS Khurmi & JK Gupta, Eurasia Publishing House(Pvt) Ltd

INDEX

C

D

E

F

G

H

I

J

K

L

M

N

O

P

Q

R

S

T

U

V

W

Y

ISBN 141200241-9

9 781412 002417